Analysis

Other titles in this series

Modular Mathematics Series

Analysis

P E Kopp

School of Mathematics
University of Hull

ELSEVIER

Elsevier Ltd
Linacre House, Jordan Hill, Oxford OX2 8DP
200 Wheeler Road, Burlington, MA 01803
Transferred to digital printing 2004
© 1996 P E Kopp

British Library Cataloguing in Publication Data
A catalogue record for this book is available from the British Library

ISBN 0 340 64596 2

Typeset in 10/12 Times by
Paston Press Ltd, Loddon, Norfolk

Contents

Series Preface

This series is designed particularly, but not exclusively, for students reading degree programmes based on semester-long modules. Each text will cover the essential core of an area of mathematics and lay the foundation for further study in that area. Some texts may include more material than can be comfortably covered in a single module, the intention there being that the topics to be studied can be selected to meet the needs of the student. Historical contexts, real life situations, and linkages with other areas of mathematics and more advanced topics are included. Traditional worked examples and exercises are augmented by more open-ended exercises and tutorial problems suitable for group work or self-study. Where appropriate, the use of computer packages is encouraged. The first level texts assume only the A-level core curriculum.

Professor Chris D. Collinson
Dr Johnston Anderson
Mr Peter Holmes

Preface

In the past decade, 'friendly' introductions to Mathematical Analysis have appeared with increasing regularity, arguably in response to the perception that the rigour and abstraction demanded by the subject no longer feature strongly in school syllabi. Most mathematicians have little doubt that, despite its beauty and elegance (or even because of these?) Analysis is among the more challenging topics faced by new undergraduates. It is certainly all too easy to take it too far too quickly!

I hope that, despite a few indulgences in Tutorial Problems and occasionally in the Exercises, this will not be the verdict on the current text. I have tried to build on the basic concepts provided by a careful discussion of *convergence*, and have stuck resolutely to sequences where possible. The text should be seen as following the foundations laid by Keith Hirst's *Numbers, Sequences and Series*, but is also designed to be fairly self-contained.

Elementary results and examples regarding convergence are reviewed in Chapter 2, while the consequences of the completeness axiom for the real numbers are developed in more detail in Chapter 3. Similarly, power series are used to provide a bridge between these techniques and the elementary transcendental functions in Chapter 4.

The main part of the book concerns the central topics of *continuity*, *differentiation* and *integration* of real functions. Here, too, sequential techniques are exploited where possible, and attention is given to the systematic approximation of complex structures by simpler ones. The 'dynamic' features of a sequential approach to topological ideas, which are so evident in the historical development of Analysis, make the subject more intuitive and accessible to the new student in my view.

Throughout, I have tried to highlight the historical context in which the subject has developed (not all that long ago!) and have paid attention to showing how increasing precision allows us to refine our geometric intuition. The intention is to stimulate the reader to reflect on the underlying concepts and ideas, rather than to 'learn' definitions which appear to come out of the blue. This means that the 'shortest route' to a major theorem is not always the most illuminating one, but the detours seem worthwhile if they increase understanding.

Much of the text is based on a one-semester module given to first-year students at Hull, who would typically have had some exposure to sequences and series already. The rewards in terms of student enlightenment are seldom immediate, but they are always enjoyable when they appear!

Acknowledgements

I am indebted to Peter Beckett for supplying the Pascal programs, the essential parts of which appear in the Appendix and which I have inexpertly tried out on my classes. Many thanks to Neil Gordon for producing the Appendix in its current form and for his expert help in creating the figures and diagrams which appear in the text. Thanks also to Marek Capinski for his patient explanation of the subtleties of LaTex, which I hope to master one day, and to David Ross of Arnold for his help and encouragement. Above all I want to thank my wife, Heather, for her unfailing support and patience. This book is dedicated to her.

Ekkehard Kopp
Hull, February 1996

1 • Introduction: Why We Study Analysis

Why study Analysis? Or better still: why prove anything? The question is a serious one, and deserves a careful answer. When computer graphics can illustrate the behaviour of even very complicated functions much more precisely than we can draw them ourselves, when chaotic motion can be studied in great detail, and seems to model many physical phenomena quite adequately, why do we still insist on producing 'solid foundations' for the Calculus? Its methods may have been controversial in the seventeenth century, when they were first introduced by Newton and Leibniz, but now it is surely an accepted part of advanced school mathematics and nothing further needs to be said? So, when the stated purpose of most introductory Analysis modules is to 'justify' the operations of the Calculus, students frequently wonder what all the fuss is about – especially when they find that familiar material is presented in what seem to be very abstract definitions, theorems and proofs, and familiar ideas are described in a highly unfamiliar way, all apparently designed to confuse and undermine what they already know!

1.1 What the computer cannot see ...

Before you shut the book and file its contents under 'useless pedantry', let's reflect a little on what we *know* about the Calculus, what its operations are and on what sort of objects it operates. If we agree (without getting into arguments about definitions) that differentiation and integration are about *functions* defined on certain *sets of real numbers*, the problems are already simple to state: just what *are* the 'real numbers' and what should we demand of a 'function' before we can differentiate it? We may decide that the numbers should 'lie on the number line' and that the functions concerned should at least 'tend to a limit' when the points we are looking at (on the x-axis, say) 'approach' a particular point a on that line. And we may require even more.

Our basic problem is that these ideas involve *infinite sets* of numbers in a very fundamental way. We can represent the idea of *convergence*, for example, as a two-person game which in principle has infinitely many stages. In its simplest form, when we want to say that an infinite sequence (a_n) of real numbers has limit a, we have two players: player 1 provides an estimate of closeness, that is, she insists that the *distance* between a_n and a should be small enough; while player 2 then has to come up with a 'stage' in the sequence beyond which *all* the a_n will satisfy player 1's requirements. To make this more precise: if player 1 nominates an 'error bound' $\varepsilon > 0$ then player 2 has to find a positive integer N such that for *all greater* integers n the distance between a_n and a is less than ε. Only then has player 2 successfully survived that phase of the game. But now player 1 has infinitely many further attempts available: she can now nominate a different error bound ε and player 2 again needs to find a suitable N, over and over again. So player 2 wins if he can *always* find a suitable N, whatever choices of ε player 1 stipulates; otherwise player 1 will win.

This imaginary game captures the spirit of convergence: however small the given error bound ε, it must always be possible to satisfy it from some point onwards. This idea cannot easily be translated into something a computer can check! Computers can check case after case, very quickly, and this can provide useful information, but they cannot provide a *proof* that the conditions will always be met.

Another simple example of this comes when we want to show that there are infinitely many prime numbers. The computer cannot check them all, precisely because there are infinitely many. However, a *proof* of this fact has been known for thousands of years, and is recorded, for example, in the Greek mathematician Euclid's famous *Elements of Geometry*, written about 300 BC. The idea is simple enough: if there were only finitely many, there would be a largest prime, p, say. But then it is not hard to show that (using the factorial $p! = 1.2.3\ldots(p-1).p$) the number $K = p! + 1$ is also prime, and is bigger than p. Thus p can't be the largest prime, and so the claim that such a prime exists leads us to a contradiction. Hence there must be infinitely many primes.

Here, as so often in mathematical proofs, the *logical sequence* of our statements is crucial, and enables us to make assertions about infinite sets, even though we are quite unable to verify each possible case separately in turn. While computers can be taught the latter, the *analytical skills* and the ability to handle *abstract concepts* inherent in such reasoning still have the dominant role in mathematics today.

This is also the stated aim of mathematics at A-level in the UK: one of the 'compulsory assessment objectives' in A-level Mathematics states that students should be able to: *construct a proof or a mathematical argument through an appropriate use of precise statements, logical deduction and inference and by the manipulation of mathematical expressions*. This book is not a political tract, so we shall refrain from commenting on how far this aim is achieved in practice.

This book is written for university students who, by and large, will have had some exposure to mathematical proof and logical deduction. Our main aim is to provide a body of closely argued material on which these skills can be honed, in readiness for the higher levels of abstraction that will follow in later years of your undergraduate course. This process requires patience, effort and perseverance, but the skills you should gain will pay off handsomely in the end.

1.2 From counting to complex numbers

First of all, what *is* a real number? We need to decide this; otherwise we can't hope to talk about *sequences* of real numbers, *functions* which take real numbers to real numbers, etc. The pictorial representation as all numbers on an infinite line is a useful guide, but hardly an adequate *definition*. In fact, through the centuries there have been many views of what this *continuum* really represents: is a line just a 'collection of dots' spaced infinitely closely together, or is it an indivisible whole, so that, however small the pieces into which we cut it, each piece is again a 'little line'? These competing points of view lead to very different perceptions of mathematics.

But let us start further back. What do we need sets of numbers for? One plausible view is that we can start with the set $\mathbb{N}$ of natural numbers as given; in the nineteenth century German mathematician Leopold Kronecker's famous phrase: *God created the natural numbers; all the rest is the work of Man.* We are quickly led to the set $\mathbb{Z}$ of all (positive and negative) integers, since positive integers allow addition, but not (always) subtraction. Within $\mathbb{Z}$ we can multiply numbers happily, but we cannot always divide them by each other (except by 1). Thus we consider ratios of integers, and create the set $\mathbb{Q}$ of all rational numbers, in which all four operations of arithmetic are possible, and keep us within the set. However, now we find – as the followers of the Greek philosopher and mathematician Pythagoras discovered to their evident dismay around 450 BC – that *square roots* present a problem: the set $\mathbb{Q}$ is found to have 'gaps' in it, for example where the 'number' $\sqrt{2}$ should be. Plugging this gap took rather a long time, and led to many detours on the way: a modern description of this journey takes up Chapters 2–5 of the companion text *Numbers, Sequences and Series* by Keith Hirst. (Historically, the *axiomatic* approach which is now taken to these problems is a recent phenomenon, even though the Ancient Greeks introduced the axiomatic method into geometry well over 2000 years ago.)

And even then mathematicians were not satisfied, since, although the equation $x^2 - 2 = 0$ could now be solved, and its solutions, $x = \sqrt{2}$, and $x = -\sqrt{2}$ made sense as members of the set $\mathbb{R}$ of real numbers, the solutions of $x^2 + 1 = 0$ did not! The final step, to the system $\mathbb{C}$ of complex numbers, occupies Chapter 6 of Hirst's book – here, however, we shall stick to the set $\mathbb{R}$ of real numbers for our Analysis.

1.3 From infinitesimals to limits

Our main concern is not with the properties of the set $\mathbb{R}$ as a single entity, but rather with the way in which its elements relate to each other. Thus we shall take the algebraic and order properties of $\mathbb{R}$ for granted, and focus on the consequences of the claim that $\mathbb{R}$ 'has no gaps'.

Just what this means bothered the Greek mathematicians considerably. The idea that lines and curves are 'made up' of dots, or even of 'infinitely short lines' is quite appealing: it allowed mathematicians to imagine that, by adding points 'one by one' to a line segment they could measure *infinitesimal* increases or decreases in its length. The known properties of regular rectilinear bodies could then be transferred to more complicated curvilinear ones. A circle, for example, could be imagined as a polygon with infinitely many infinitely short sides: which leads to a simple proof of the area formula: imagine the circle of radius r as made up of infinitely many infinitely thin isosceles triangles, each with height infinitely close to r and infinitesimal base b. Each has area equal $\frac{1}{2}br$, so the area of the circle is $A = \frac{1}{2}rC$, where C (the sum of all the bases b) is the circumference of the circle. But if π is the ratio of circumference to diameter, we also have $C = 2\pi r$, so that, substituting for C, we obtain $A = \pi r^2$.

Though the logical difficulties of adding infinitely many quantities (while their sum remained finite) and dividing finite quantities by infinitesimals soon discredited such techniques, they have stayed with mathematicians throughout the centuries as useful heuristic devices. They flourished again in build-up to the

Calculus in the sixteenth and seventeenth centuries: Johann Kepler, for example, gave a three-dimensional version of the above argument, showing how the sphere is 'made up' of infinitely thin cones and hence that its volume V is $\frac{r}{3}$ times the surface area ($A = 4\pi r^2$), yielding $V = \frac{4}{3}\pi r^3$. Gottfried Wilhelm Leibniz, in particular, sought to put the infinitesimals dx on a proper logical footing in order to justify statements like

$$\frac{(x + dx)^2 - x^2}{dx} = 2x + dx \approx 2x$$

where the symbol $\approx$ denotes that the quantities differ only by an infinitesimal amount. Much effort was expended to resolve the paradox involved in first dividing by the quantity dx and then ignoring it as if it were 0, and throughout the eighteenth and early nineteenth century this led to a gradual realization of how we could describe these ideas using *limits*, and that a proper analysis of the classes of *functions* which describe the curves involved should precede any justification of the Calculus. This led away from pictorial representation, and a closer look at the *number systems* on which such functions had to be defined. The wheel had now turned full circle, and by the early nineteenth century mathematicians could begin their study of the newly independent subject of Analysis.

2 • Convergent Sequences and Series

This chapter is devoted to a self-contained *review* of the properties of convergent sequences and series, which are described in more detail in the companion text *Numbers, Sequences and Series* by Keith Hirst, Chapters 7–9. This text will henceforth be referred to as [*NSS*]. If you have no previous experience of the fundamental idea of *convergence of a real sequence*, or wish to refresh your memory, you should consult this text and practise your skills on the examples and exercises provided there, which complement those presented in this book. The idea of convergence is a fundamental theme of the present book, and the results discussed in this chapter will be used throughout those that follow. The definitions of the number systems $\mathbb{N}$, $\mathbb{Z}$, $\mathbb{Q}$, $\mathbb{R}$ and $\mathbb{C}$ will be taken for granted in this book: details of these can also be found in [*NSS*].

The terms 'sequence' and 'series' are often used interchangeably in ordinary language. This is a pity, since the distinction between them is very simple, and yet very useful and important. We shall take care *not* to confuse them.

2.1 Convergence and summation

Sequences are 'lists' of numbers, often generated by an inductive procedure, such as those familiar from early number games in which we have to 'guess' the next number. For example, given 1,3,6,10,15,... we might decide that the next number should be 21, since the difference between successive numbers increases by 1 each time. We either need to be given sufficiently many terms to deduce the rule of succession, or we can be given the rule directly.

◈ Example 1

The above sequence of *triangular numbers* is obtained by writing $a_1 = 1$, $a_2 = 1 + 2 = a_1 + 2$, $a_3 = a_2 + 3$, etc. The reason for the name becomes clear when we represent each unit by a dot in the following pattern:

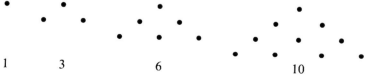

1   3   6   10

The properties of this sequence were well known to the Ancient Greeks. Pythagoras and his followers studied them in some detail, for example. For each $n \geq 1$, $a_n - a_{n-1} = n$. Therefore we can find the value of a_n by starting at $a_1 = 1$ and successively adding in all the differences, i.e. $a_n = 1 + 2 + 3 + \ldots + n$. The sum is easy to calculate:

$$a_n = 1 + 2 + \ldots + (n - 1) + n$$
$$a_n = n + (n - 1) + \ldots + 2 + 1$$

Adding, we have $2a_n = n(n + 1)$, so we have proved the well-known formula

$$1 + 2 + 3 + \ldots + n = \frac{1}{2}n(n + 1)$$

In this example the inductive definition reduces to a *formula* which allows us to express a_n as a *function* of n, namely $a_n = \frac{1}{2}n(n + 1)$. There are many such examples.

Pythagoras (*ca.* 580–495 BC) and his followers were perhaps the first group to attempt an explanation of their physical environment in quantitative terms, i.e. associating numbers with objects. Whole numbers served as the building blocks of the universe for them, and they were thus led into a study of the relationship between numbers, based in part on figurative numbers such as those shown above.

◉ *Example 2*

The sequence $\{1, 4, 9, 16, \ldots\}$, all consisting of perfect squares, is fully described by writing $\{n^2 : n \geq 1\}$; similarly we have two descriptions of the sequence $\{2^{-n} : n \geq 1\} = \{\frac{1}{2}, \frac{1}{4}, \frac{1}{8}, \frac{1}{16}, \ldots\}$. In each case $a_n = f(n)$ for some function f.

In these examples the sequences generated are (at least potentially) *infinite*: they do not stop, and for each $n \in \mathbb{N}$ we can find the nth *term* a_n of the sequence (a_n). In our examples these terms are also real numbers; hence a *real sequence* $(a_n)_{n \geq 1}$ is thought of informally as an infinite list of real numbers, usually given by some rule for forming each term in the list.

All *infinite* sequences have one thing in common: to each $n \in \mathbb{N}$ there corresponds a uniquely defined real number a_n. We turn this into a quite general *definition*:

A **real sequence** is a function $n \mapsto a_n$ with domain $\mathbb{N}$ and with range contained in $\mathbb{R}$.

Of course, we have not formally defined the term 'function' so far, and without this the above definition is not very meaningful. We shall give formal definitions in Chapter 5; for the present the simple idea of a function as a 'rule' suffices to make clear that we wish to associate a real number a_n with *each* $n \in \mathbb{N}$.

However, the terms a_m and a_n need not be different if $m \neq n$: for example, if $a_n = (-1)^n$, then every even-numbered term is 1 and each odd-numbered term is -1. Here the range of the function $n \mapsto a_n$ is just the two-point set $\{-1, 1\}$. Similarly, we can define *constant* sequences simply by setting $a_n = c$ (where c is a fixed real number) for all $n \in \mathbb{N}$.

The principal purpose of our analysis of infinite sequences is to provide a convenient tool for *approximation* and *convergence*. For example, we can approach the irrational number $\sqrt{2}$ by writing down successive decimal approximations, such as 1.4, 1.41, 1.414, 1.4142, 1.41421, 1.414213, 1.4142136, etc. Since $\sqrt{2}$ cannot be written as a finite decimal, the approximating sequence will never 'reach' the target value $\sqrt{2}$, but by taking enough terms we can approximate it to as high a degree of accuracy as we wish.

Despite our formal definition we usually think of the terms of sequence (a_n) as values which are 'taken in succession' as n increases. This is especially useful in dealing with sequences which are defined *recursively*: if, for each x_n, its *successor* (the next term) x_{n+1} is defined as some function of x_n (and possibly some of its earlier *predecessors*, $x_{n-1}, x_{n-2}, \ldots, x_1$) then we may be able to guess what happens to x_n 'in the long run'.

● *Example 3*

Set $x_1 = 1$, and for each n, suppose that $x_{n+1} = \frac{x_n^2+3}{2x_n}$. Here it is no longer clear how we might write $x_n = f(n)$ for some explicit function f, but we have $x_2 = \frac{4}{2} = 2$, $x_3 = \frac{7}{4} = 1.75$, and similarly $x_4 = 1.732143$, $x_5 = 1.732051$, etc. So it seems that this sequence 'settles down' rather quickly at $\sqrt{3}$. We can 'check' this by writing the 'limit' of the sequence as x, so that x *should* be a solution of the equation $x = \frac{x^2+3}{2x}$ (why?), and this obviously simplifies to $x^2 = 3$.

In fact, even the attempt to *define* irrationals like $\sqrt{3}$ via their decimal representation requires us to specify precisely what we mean by the 'limiting value' of a sequence: we could consider the sequence $(x_n)_{n\geq1}$ whose terms are given successively as $x_1 = 1$, $x_2 = 1.7$, $x_3 = 1.73$, $x_4 = 1.732$, etc. – and then the requirement is to make sense of 'x_∞' (!) We shall solve this problem shortly.

TUTORIAL PROBLEM 2.1

The use of *iteration* to provide approximate solutions to equations can be taken much further: read [*NSS*] Chapter 7.2 for more examples and test your skills on the Exercises at the end of that section. We shall return to these problems in Chapter 7, when we will be able to justify the techniques used much more fully.

Series are, informally, 'what you get when you add together the terms of a sequence'. But we cannot add infinitely many numbers together directly, so this vague claim needs to be made more precise. What we do, given the terms of a sequence $(a_i)_{i\geq1}$, is to form the *partial sums* $s_1 = a_1, s_2 = a_1 + a_2, s_3 = a_1 + a_2 + a_3$, etc., and in general:

$$s_n = a_1 + a_2 + a_3 + \ldots + a_n = \sum_{i=1}^{n} a_i$$

For each $n \geq 1$ we call s_n the nth *partial sum* of the *series* $\sum_i a_i$. Strictly speaking, the series is itself defined by this sequence of partial sums, so that a formal definition of a series involves listing the sequence of its terms (a_i) as well as the sequence (s_n) of partial sums – and each can be found from the other (*see* [*NSS*], Definition 8.1 for a more formal description).

We shall reserve the notation $\sum_i a_i$ for the series whose terms are given by members of the sequence (a_i).

Of course, this still doesn't answer the question of what we should mean by the *sum* of the series $\sum_i a_i$. Fortunately, we can decide this as soon as we know how to define the 'limiting value' 's_∞' of the sequence of its partial sums, so that we handle both our problems at once.

TUTORIAL PROBLEM 2.2

As an alternative to the usual 'Achilles and the Tortoise' example (*see* [*NSS*], Chapter 8), imagine a snail crawling along infinitely expandable rubber bands in various ways. In each case we are given that the snail crawls 1 m each day, and that it rests at night. While it rests, a demon secretly stretches the rubber band. In each of the following examples the question is: does the snail (which starts from one end of the band) ever reach the other end, and, if so, how long does it take?

  (i) The initial length of the rubber band is 10 m. Every night the demon stretches it by a further 10 m.
 (ii) The initial length of the rubber band is 4 m. Every night the demon doubles its previous length.
(iii) This time the rubber band is initially 2 m long, and the demon doubles the length every night.
 (iv) The rubber band is x metres long at the start, and the demon increases its length by 25% every night. What is the 'critical' value x^* so that for all $x < x^*$ the snail will reach the other end in a finite time? Given that $x^* > 4$, find how long the snail will take when $x = 4$.

Limits of infinite sequences

The above examples all have one thing in common: we are interested in what happens 'eventually' or 'in the long run' as we move down our sequence (a_n) – or along the corresponding sequence of partial sums. If we want the sequence to 'have a limit x', then its terms will 'eventually' have to be 'as close as we please' to x. In other words, the distance between x_n and x becomes *arbitrarily small* provided we take n *sufficiently large*. This leads to the following fundamental concept:

● *Definition I*

  (i) The real sequence (x_n) *converges to the finite limit x as n tends to infinity* (written $x_n \to x$ as $n \to \infty$, or $x = \lim_{n\to\infty} x_n$) if: for every real $\varepsilon > 0$ there exists an $N \in \mathbb{N}$ such that $|x_n - x| < \varepsilon$ whenever $n > N$.
 (ii) If (x_n) has no finite limit, we say that the sequence *diverges*.

Formally, part (ii) includes the case when (x_n) 'goes to infinity': for example, if for every given $K > 0$ we can find $N \in \mathbb{N}$ such that $x_n > K$ whenever $n > N$.

By a slight abuse of notation we shall nevertheless write $x_n \to \infty$ when this happens. A similar interpretation is given to the notation $x_n \to -\infty$. Rather confusingly, some authors describe this by saying that '(x_n) diverges to ∞' (or $-\infty$).

In order to apply this definition to some useful examples we need just one property of the real line $\mathbb{R}$ at this stage – that $\mathbb{R}$ contains 'arbitrarily large' natural numbers. We formulate this, without proof, as the:

● *Principle of Archimedes*

Given any real number x there exists $n \in \mathbb{N}$ such that $n > x$.

With a precise definition of $\mathbb{R}$ via a set of *axioms* one can prove this (rather obvious?) Principle rigorously – we shall show later that it follows easily from the *Completeness Axiom* for $\mathbb{R}$. These issues are discussed fairly fully in [*NSS*], Chapter 5 – *see* Proposition 5.6. We shall explore the consequences of the Completeness Axiom for the convergence of sequences and series in more detail in Chapter 3.

Example 4

Let $x_n = \frac{1}{n}$ for each $n \geq 1$. This is the sequence $1, \frac{1}{2}, \frac{1}{3}, \ldots, \frac{1}{n}, \ldots$ which should clearly have limit 0. To prove this, suppose some $\varepsilon > 0$ is given. How do we find an 'appropriate' N? Now if $n > N$ then $\frac{1}{n} < \frac{1}{N}$, hence we only need to find a *single* N such that $\frac{1}{N} < \varepsilon$, which is the same as demanding that $N > \frac{1}{\varepsilon}$. But since $\frac{1}{\varepsilon}$ is a real number, the Principle of Archimedes ensures that such an N must exist in $\mathbb{N}$. Therefore we have proved that $\frac{1}{n} \to 0$ when $n \to \infty$. Note that the N we found above need *not* be the first integer greater than $\frac{1}{\varepsilon}$; *any* such integer will do! However, in general the number N will depend on the given ε, and so to prove convergence we need to supply a 'rule' which chooses $N = N(\varepsilon)$, and which does so for *every* given $\varepsilon > 0$.

Example 5

$x_n = \frac{n}{n+1}$ is the sequence $\frac{1}{2}, \frac{2}{3}, \frac{3}{4}, \ldots$ and it seems clear that its limit should be 1. To *prove* this we need to consider the *distance* $|x_n - 1| = 1 - \frac{n}{n+1} = \frac{1}{n+1}$ and since this clearly converges to 0 as $n \to \infty$ it follows at once that $\lim_{n \to \infty} x_n = 1$.

Example 6

On the other hand, the sequence $-1, 1, -1, 1, -1, 1, -1, 1, \ldots$, which we express conveniently as $x_n = (-1)^n$, seems never to settle down, and the difference between consecutive terms is always 2, since the values of the terms *oscillate* forever between -1 and 1.

To prove that (x_n) *diverges*, note first that $|x_n - x_{n+1}| = 2$ for all n, as observed above. Now if the sequence (x_n) had a limit x, say, then $\lim_{n \to \infty} x_{n+1} = x$ also, by the definition of convergence. So we would be able to find N' and N'' such that $|x_n - x| < 1$ for $n > N'$, and $|x - x_{n+1}| < 1$ for $n > N''$. Hence for all $n > N = \max(N', N'')$ both these inequalities must hold. But this contradicts the fact that $|x_n - x_{n+1}| = 2$.

To see this we first need to recall a basic inequality which is left as an exercise (or: *see* [*NSS*], Proposition 6.5).

TUTORIAL PROBLEM 2.3

(i) Prove the *Triangle Inequality* for real numbers: For any real numbers a, b, $|a + b| \leq |a| + |b|$.

(ii) Deduce that for any real numbers a, b, $||a| - |b|| \leq |a - b|$.

Back to our sequence: $x_n = (-1)^n$. Applying the Triangle Inequality to $a = x_n - x$, $b = x - x_{n+1}$, we obtain

$$|x_n - x_{n+1}| = |(x_n - x) + (x - x_{n+1})| \leq |x_n - x| + |x - x_{n+1}|$$

But the distance on the left is 2, while the sum on the right is *less* than $1 + 1 = 2$. The contradiction arises from our assumption that $x = \lim_{n \to \infty} x_n$ exists, hence this limit *cannot* exist. Therefore the sequence (x_n) diverges.

On the other hand, this sequence is *bounded*: by this we mean that there exists $K \in \mathbb{R}$ such that $|x_n| \leq K$ for all $n \geq 1$. In fact in the present example we even have $|x_n| \leq 1$ for all n.

So: a bounded sequence need not always have a limit.

Example 7

Many sequences are *unbounded*: there is *no* number K such that $|x_n| \leq K$ for all $n \in \mathbb{N}$. In other words: given any $K > 0$ we can always find n such that $|x_n| > K$. For example:

(i) $1, 2, 3, 4, \ldots, n, \ldots$ (i.e. $x_n = n$);

(ii) $1, 2, 4, 8, \ldots (x_n = 2^n)$;

(iii) $x_n = (-n)^n$;

(iv) the famous *Fibonacci sequence* (x_n) is defined inductively by: $x_0 = 0$, $x_1 = 1$, $x_n = x_{n-1} + x_{n-2}$ for $n \geq 2$. In other words: $(x_n) = 0, 1, 1, 2, 3, 5, 8, 13, 21, 34, \ldots$.

That all four sequences above *diverge* follows from a simple fact:

● Proposition 1

Every convergent sequence is bounded. Hence every unbounded sequence diverges.

PROOF

Suppose that $x_n \to x$ as $n \to \infty$. Then we can find N such that $|x_n - x| < 1$ for all $n > N$. Since the Triangle Inequality yields

$$||x_n| - |x|| \leq |x_n - x|$$

it follows that $|x_n| < 1 + |x|$ for all $n > N$. Therefore we have shown that $K = \max\{|x_1|, |x_2|, \ldots |x_N|, 1 + |x|\}$ satisfies $|x_n| \leq K$ for all $n \geq 1$. In other words, (x_n) is bounded by $-K$ and K.

Thus the condition '(x_n) is bounded' is *necessary* for convergence of (x_n). On the other hand, Example 6 shows that it is *not sufficient*, since $x_n = (-1)^n$ defines a sequence which is bounded but divergent.

So far we have invariably used an educated guess to find the limit, and then proved that our guess was correct. But how do we know that our guess yields the *only* possible limit of (x_n) – in other words, is the limit *unique* when it exists?

To prove that it is unique, suppose that the sequence (x_n) converges to both x and y. We show that then $x = y$. To do this, we need only show that the *distance* $|x - y|$ is less than every number $\varepsilon > 0$. (Think about this!) But if $\varepsilon > 0$ is fixed, then we can find N such that $|x - x_n| < \frac{\varepsilon}{2}$ for $n > N$, and N' such that $|x_n - y| < \frac{\varepsilon}{2}$ for $n > N'$. Using the Triangle Inequality again, we obtain, whenever $n > \max(N, N')$:

$$|x - y| = |(x - x_n) + (x_n - y)| \leq |x - x_n| + |x_n - y| < \varepsilon$$

Since ε was arbitrary, it follows that $x = y$.

Thus: the limit of a convergent sequence is unique.

TUTORIAL PROBLEM 2.4

As an exercise in induction, prove an inequality which will become useful in much of what follows:

Bernoulli's Inequality: for $x > -1$ and $n \in \mathbb{N}$ we have $(1 + x)^n \geq 1 + nx$.

Discuss why we need the condition $x > -1$. Is there a more direct proof if we restrict to $x \geq 0$?

Sums of infinite series

We now consider the special case when the sequence in question is regarded as the sequence (s_n) of *partial sums* $s_n = \sum_{i=1}^{n} a_i$ of a *series* $\sum_i a_i$.

● *Definition 2*

We say that the *series* $\sum_i a_i$ of real numbers *converges* if the sequence (s_n) of its partial sums converges to a (finite) limit, s. In that case we call s the *sum of the series* and write $s = \sum_{i=1}^{\infty} a_i$.

In other words, $\sum_{i=1}^{\infty} a_i$ is by definition equal to $s = \lim_{n \to \infty} s_n = \lim_{n \to \infty} \sum_{i=1}^{n} a_i$ (which explains the notation!) and the symbol $\sum_{i=1}^{\infty} a_i$ therefore denotes the *real number s*, while we shall consistently write $\sum_i a_i$ for the series itself (which is, as we have seen, determined by its sequence (s_n) of partial sums).

Initially we shall usually deal with examples where all *terms* a_i of the series are *positive*. In that case, the sequence (s_n) of partial sums is *increasing*, since $s_n - s_{n-1} = a_n > 0$. This greatly simplifies matters.

Example 8

Consider the *geometric series*: $\sum_{i\geq 0} x^i$ where $x \in \mathbb{R}$ is fixed. (By convention we start this series with $a_0 = 1 = x^0$, rather than with a_1.) First, if $x = 1$, we obtain

$$s_n = \sum_{i=0}^{n} 1^i = 1 + 1 + 1 + 1 + \ldots + 1 = n + 1$$

which is obviously unbounded above (*again* using the Archimedes Principle!); for $x = -1$ we obtain $s_n = 1 - 1 + 1 - 1 + 1 \ldots \pm 1$, which is either 0 or 1, depending on whether n is even or odd. Hence again (s_n) diverges. For $|x| > 1$ the sequence (s_n) also diverges – this can be shown similarly, and also follows at once from Proposition 2 below.

Hence we need only consider the case $|x| < 1$. But here the partial sums are easy to find, since $x \neq 1$:

$$(1 - x) \sum_{i=0}^{n} x^i = (1 - x)(1 + x + x^2 + x^3 + \ldots + x^n) = 1 - x^{n+1}$$

hence $s_n = \frac{1 - x^{n+1}}{1 - x}$ for each n.

Our claim is that $s_n \to \frac{1}{1-x}$ as $n \to \infty$: to prove this, it will suffice to prove that, when $|x| < 1$, then $x^n \to 0$ as $n \to \infty$ – so that we are back to a simple result about sequences! This result is also proved in [*NSS*], Proposition 7.3, but for a bit of variety we include a different proof here.

Write $a = |x|$, then $0 \leq a < 1$. It will clearly be enough to show that $a^n \to 0$ as $n \to \infty$. The case $a = 0$ is obvious, since (a^n) is the constant sequence $\{0, 0, 0, 0, \ldots\}$ which converges to 0. For $0 < a < 1$ we have $\frac{1}{a} = 1 + h > 1$, and hence

$$\frac{1}{a^n} = \left(\frac{1}{a}\right)^n = (1 + h)^n \geq 1 + nh$$

is unbounded above (we have used the Bernoulli Inequality; *see* Tutorial Problem 2.4). Thus if $\varepsilon > 0$ is given, we can find N such that for $n > N$, $\frac{1}{a^n} > \frac{1}{\varepsilon}$, i.e. $a^n < \varepsilon$. Thus: for $|x| < 1$, $x^n \to 0$ as $n \to \infty$.

Hence, finally: the sum of the geometric series $\sum_{i\geq 0} x^i$ is $\frac{1}{1-x}$ whenever $|x| < 1$. For all other $x \in \mathbb{R}$ the sum is undefined, i.e. the series diverges.

In particular you should now recognize the solution of the *snail problem* (Tutorial Problem 2.2) in each of the cases when the demon *doubled* the rubber band each time, since we then had to sum the series $\sum_i \frac{1}{2^i}$. We start from $i = 2$ when the band is 4 m long at the beginning – hence leading to the sum $\frac{1}{4}(\frac{1}{1 - \frac{1}{2}}) = \frac{1}{2}$, which means that the snail never gets beyond halfway – and from $i = 1$ for the 2 m band, so that the final sum is 1, and the snail reaches the other end 'in the limit' only! These are therefore special cases of the geometric series.

Example 9—The harmonic series

This is the series of reciprocals $\sum_n \frac{1}{n}$, i.e. $1 + \frac{1}{2} + \frac{1}{3} \ldots + \frac{1}{n} + \ldots$ We shall show that this series *diverges*, because the sequence (s_n) of its partial sums is unbounded above. This provides the reason why the original snail in our problem *does* manage to reach the other end (with perseverance) – we needed to show that the partial sums will be greater than 10 for some n. The proof that (s_n) is unbounded above is a useful exercise in induction – we shall show that for all $n \in \mathbb{N}$,

$$1 + \frac{1}{2} + \frac{1}{3} \ldots + \frac{1}{2^n} \geq 1 + \frac{n}{2}$$

When $n = 1$ the inequality is clearly true, as both sides equal $\frac{3}{2}$. Suppose that the inequality holds for $n = k - 1$, that is: $1 + \frac{1}{2} + \frac{1}{3} \ldots + \frac{1}{2^{k-1}} \geq 1 + \frac{k-1}{2}$. We have to show that with this assumption the inequality also holds for $n = k$. So consider

$$1 + \frac{1}{2} + \frac{1}{3} \ldots + \frac{1}{2^k} = \left(1 + \frac{1}{2} + \frac{1}{3} \ldots + \frac{1}{2^{k-1}} \right) + \left(\frac{1}{2^{k-1} + 1} + \frac{1}{2^{k-1} + 2} + \ldots + \frac{1}{2^k} \right)$$

The first bracket on the right is at least $1 + \frac{k-1}{2}$ by our assumption, and the second bracket consists of 2^{k-1} terms, each of which is at least $\frac{1}{2^k}$. Hence the second bracket is greater than $\frac{1}{2}$ and the whole sum is at least $1 + \frac{k}{2}$, which proves that the inequality holds for all n.

Our inequality proves that (s_n) is unbounded above, since if there were $K > 0$ such that $s_n \leq K$ for all $n \in \mathbb{N}$ then we would need $K > 1 + \frac{n}{2}$ for all $n \in \mathbb{N}$, which contradicts the Archimedes Principle. Hence: the harmonic series diverges.

This example provides still more: it is possible for a series to diverge *even though* the sequence (a_n) of its *terms* converges to 0: in the case of the harmonic series the terms $a_n = \frac{1}{n}$ converge to 0, but not 'fast enough' to ensure that their sum is finite. On the other hand we have:

• *Proposition 2*

If the series $\sum_i a_i$ converges, then $a_n \to 0$ as $n \to \infty$.

PROOF
The partial sums (s_n) converge to the sum s of the series, and $a_n = s_n - s_{n-1}$; hence $\lim_{n \to \infty} a_n = \lim_{n \to \infty}(s_n - s_{n-1}) = \lim_{n \to \infty} s_n - \lim_{n \to \infty} s_{n-1} = s - s = 0$.

Thus, in order to show that a given series $\sum_i a_i$ *diverges* we need only show that the terms a_n do *not* converge to 0. On the other hand, simply checking that they *do* converge to 0 is not enough to ensure convergence of $\sum_i a_i$.

EXERCISES ON 2.1

1. Find the limit of the following sequences:
   (i) $x_n = \sqrt{n+1} - \sqrt{n}$ (ii) $x_n = \log_e(n+1) - \log_e n$
2. Why are the following series divergent?
   (i) $\sum_{n \geq 1}(\sqrt{n+1} - \sqrt{n})$ (ii) $\sum_{n \geq 1} \cos(\frac{1}{n})$.
3. Suppose that the sequence (a_n) converges to a, and each $a_n \in \mathbb{N}$. Show that $a \in \mathbb{N}$ and that for all large enough n, every a_n equals a.

2.2 Algebraic and order properties of limits

Strictly speaking, we have cheated in the proof of Proposition 2: our argument assumes that the limit of the difference of two sequences is the difference of their limits. This needs proof first.

We therefore investigate how limits interact with addition and multiplication of numbers: is it always true that if $a_n \to a$ and $b_n \to b$ when $n \to \infty$ then $a_n + b_n \to a + b$ and $a_n b_n \to ab$, and similarly for their difference and quotient? This would be convenient, since it would provide a simple method for handling sequences such as $\frac{n^3 - 3n + 7}{4n^3 + 2n^2 - 6}$, whose limit is *not* exactly 'obvious' at first sight! Once the above results are proved, however, we can simply divide top and bottom by n^3 and write

$$\lim_{n \to \infty} \frac{1 - \frac{3}{n^2} + \frac{7}{n^3}}{4 + \frac{2}{n} - \frac{6}{n^3}} = \frac{1 - \lim_{n \to \infty} \frac{3}{n^2} + \lim_{n \to \infty} \frac{7}{n^3}}{4 + \lim_{n \to \infty} \frac{2}{n} - \lim_{n \to \infty} \frac{6}{n^3}} = \frac{1}{4}$$

since each of the limits involved can be traced back to $\lim_{n \to \infty} \frac{1}{n} = 0$.

Null sequences

For simplicity, single out sequences whose limit is 0: call (x_n) a *null sequence* (or simply *null*) if $x_n \to 0$ as $n \to \infty$. Clearly for any sequence (a_n) we have: $a_n \to a$ as $n \to \infty$ if and only if (x_n) is null, where $x_n = a_n - a$ for each $n \in \mathbb{N}$.

The following facts are straightforward:

● *Proposition 3*

Let (x_n) and (y_n) be null sequences. Then the sequences $(x_n + y_n)$ and $(x_n y_n)$ are also null sequences.

PROOF
$(x_n + y_n)$ is formed by adding corresponding terms of (x_n) and (y_n). To show that it is null, let $\varepsilon > 0$ be given, and find N_1 and N_2 respectively such that $|x_n| < \frac{\varepsilon}{2}$ for all $n > N_1$ and $|y_n| < \frac{\varepsilon}{2}$ for $n > N_2$. Thus for $n > N = \max(N_1, N_2)$ we have: $|x_n + y_n| \leq |x_n| + |y_n| < \varepsilon$. Similarly, $(x_n y_n)$ is formed by multiplying corresponding terms of (x_n) and (y_n), and if $\varepsilon > 0$ is given we can find N_1 and N_2 respectively such that $|x_n| < \sqrt{\varepsilon}$ for $n > N_1$ and $|y_n| < \sqrt{\varepsilon}$ for $n > N_2$. So again for $n > N = \max(N_1, N_2)$ we obtain $|x_n y_n| = |x_n||y_n| < \sqrt{\varepsilon}\sqrt{\varepsilon} = \varepsilon$.

Multiplying a null sequence by a bounded sequence will again produce a null sequence: to prove this, let (x_n) be null and suppose (y_n) is bounded by K. For given $\varepsilon > 0$ find N such that $|x_n| < \frac{\varepsilon}{K}$ whenever $n > N$. Then we have $|x_n y_n| < \frac{\varepsilon}{K} K = \varepsilon$ for all such n, hence $(x_n y_n)$ is null. In particular, taking (y_n) to be the *constant* sequence $\{K, K, K, ...\}$ we have shown that (Kx_n) is also null. (So, for example, $\lim_{n \to \infty} \frac{2}{n} = 0$ in the above; more generally, since each term in $(\frac{1}{n^3})$ is a product $\frac{1}{n} \cdot \frac{1}{n} \cdot \frac{1}{n}$, Proposition 3 shows that $\lim_{n \to \infty} \frac{1}{n^3} = 0$, and so $\lim_{n \to \infty} \frac{7}{n^3} = 0$ also.) Similarly, $x_n = \frac{(-1)^n}{n}$ defines a null sequence, as $-1, 1 - 1, 1, ...$ is bounded.

Limits and arithmetic

We are now in a position to prove the first main result on the algebra of limits (*see also* [*NSS*], Proposition 7.2 for an alternative proof):

● *Theorem 1*─────────────────────────

Suppose that $x_n \to x$ and $y_n \to y$. Then we have:

(i) $(x_n + y_n) \to (x + y)$ and $(x_n - y_n) \to (x - y)$
(ii) $x_n y_n \to xy$
(iii) $\frac{x_n}{y_n} \to \frac{x}{y}$ provided $y \neq 0$.

PROOF

Let $a_n = x_n + y_n - x - y$. We show that (a_n) is null, which proves the first claim: but $a_n = (x_n - x) + (y_n - y)$ and both sequences on the right are null. Thus so is their sum, by Proposition 3. Next, write $(x_n - x) - (y_n - y) = (x_n - x) + (-1)(y_n - y)$ and note that $(-1)(y_n - y)$ is also null, by our remark above. So the second claim in (i) is proved also.

For (ii), we write $(x_n y_n - xy)$ as a sum of null sequences: $x_n y_n - xy = (x_n - x)(y_n - y) + x(y_n - y) + y(x_n - x)$ and again use Proposition 3 and the remark about bounded sequences.

Finally, to prove (iii) we write $\frac{x_n}{y_n} = x_n(\frac{1}{y_n})$, so that we need only prove $\frac{1}{y_n} \to \frac{1}{y}$ and then use (ii). But first we need to be sure that $\frac{1}{y_n}$ *makes sense*, i.e. that $y_n \neq 0$. This is where we use the condition $y \neq 0$: with it, we can define $\varepsilon = \frac{|y|}{2} > 0$ and therefore find N such that $|y_n - y| < \varepsilon$ for all $n > N$. This means that $|y| - |y_n| < \frac{|y|}{2}$, so that $|y_n| > \frac{|y|}{2} > 0$. Hence, at least if we discard the first N terms, all the $y_n \neq 0$. In fact, we have shown more: $\frac{1}{|y_n|} < \frac{2}{|y|}$, so the sequence $(\frac{1}{y_n})$ is bounded. Now consider $\frac{1}{y_n} - \frac{1}{y} = (\frac{1}{y})(\frac{1}{y_n})(y_n - y)$: the first factor on the right is constant, the second has just been shown to be bounded, and the third is a null sequence. Hence the product is null, i.e. $\frac{1}{y_n} \to \frac{1}{y}$ as $n \to \infty$.

● *Example 10*

It will be useful to have a stock of basic null sequences:

(i) we already know that $x^n \to 0$ as $n \to \infty$ for $|x| < 1$. What happens to $\sqrt[n]{x} = x^{1/n}$ for positive x? We can use the binomial theorem here: first consider the case $x > 1$, so that $x^{1/n} > 1$ also. Hence we can write $x = (1 + a_n)^n$ for some positive a_n. Expanding, we obtain $x = 1 + na_n + \ldots + a_n^n$ so that $x > na_n$, i.e. $a_n < x(\frac{1}{n})$, hence (a_n) is null (any positive sequence bounded above by a null sequence is obviously null also). Therefore $\sqrt[n]{x} = 1 + a_n \to 1$ as $n \to \infty$. Finally, when $0 < x < 1$ we can apply the above to $y = \frac{1}{x}$, and note that $\sqrt[n]{x} = \frac{1}{\sqrt[n]{y}}$. Hence for all $x > 0$, $\lim_{n \to \infty} \sqrt[n]{x} = 1$.

(ii) Now extend this and consider $x_n = n^{1/n}$: then $\lim_{n \to \infty} n^{1/n} = 1$. In fact, $x_n \geq 1$ for all n, since $n^{1/n} < 1$ would yield $n < 1^n = 1$. So if (x_n) did not converge to 1, we could find $\varepsilon > 0$ for which $x_n \geq 1 + \varepsilon$ for arbitrarily large n (i.e. *no N* can be found such that $|x_n - 1| < \varepsilon$ for all $n > N$). But then $n \geq (1 + \varepsilon)^n = 1 + n\varepsilon + \frac{n(n-1)}{2}\varepsilon^2 + \ldots + \varepsilon^n > \frac{n(n-1)}{2}\varepsilon^2$, hence $\frac{2}{\varepsilon^2} > n - 1$ for arbitrarily large n, which contradicts the Archimedes Principle. Therefore $\sqrt[n]{n} = n^{1/n} \to 1$ as $n \to \infty$.

(iii) For any non-negative null sequence (x_n) it is easy to see that for all $p \in \mathbb{R}_+$, (x_n^p) is also null: given $\varepsilon > 0$ we can choose N large enough to guarantee that $|x_n| < \varepsilon^{1/p}$ for all $n \geq N$. Then $|x_n^p| < \varepsilon$ for all $n \geq N$, so $x_n^p \to 0$ as $n \to \infty$. In particular, since $\frac{1}{n} \to 0$ it follows that $\frac{1}{n^p} \to 0$ for all $p > 0$. (If the *definition* of x_n^p worries you (especially when $p \notin \mathbb{Q}$) then you should consult Chapter 6!)

Recall that convergence of a series $\sum_i a_i$ simply involves the convergence of the sequence of its partial sums (s_n), where $s_n = \sum_{i=1}^n a_i$ for each n. So Theorem 1 implies: if $\sum_i a_i$ and $\sum_i b_i$ are convergent series, with sums a and b respectively, and if α and β are real numbers, then the series $\sum_i (\alpha a_i + \beta b_i)$ converges to $\alpha a + \beta b$. To see this simply note that if (s_n) and (t_n) are the partial sums of $\sum_i a_i$ and $\sum_i b_i$ respectively, then $(\alpha s_n + \beta t_n)$ is the sequence of partial sums of $\sum_i (\alpha a_i + \beta b_i)$. Hence, for example, using Further Exercise 4(ii) below we obtain:

$$\sum_{i=0}^{\infty} \left(\frac{\pi}{5} \left(\frac{1}{2} \right)^i + \frac{3n}{(n+1)!} \right) = \frac{2\pi}{5} + 3$$

No such results work for *products*, however, since the product $\sum_{i=1}^n a_i b_i$ does *not* equal $s_n t_n$ for $n > 1$. The question of convergence of products of series is altogether more subtle and is left till Chapter 4.

Limits and inequalities

The interaction with *order properties* of the real numbers, producing *inequalities*, is our next stop: it seems clear that if $x_n \leq y_n$ for all $n \in \mathbb{N}$, then the same relation should hold for their limits. But beware: this breaks down for $x_n < y_n$, since if $x_n = 0$ and $y_n = \frac{1}{n}$ for all n, then the limits of the sequences are equal, even though none of their corresponding terms are. So there *is* something to prove here.

● *Proposition 4*

Suppose that for all n greater than some fixed n_0, $x_n \leq y_n$ and that $x_n \to x$ and $y_n \to y$ as $n \to \infty$. Then $x \leq y$.

PROOF
Suppose that $x > y$. Then $\varepsilon = \frac{x-y}{2} > 0$, so we can find N_1 and N_2 such that $x_n > x - \varepsilon$ for $n > N_1$ and $y_n < y + \varepsilon$ for $n > N_2$. But $x - \varepsilon = \frac{x+y}{2} = y + \varepsilon$. Hence for $n > N = \max(N_1, N_2, n_0)$ we have $y_n < x_n$, which contradicts $x_n \leq y_n$. Thus $x \leq y$.

(*Note* that this gives another proof of the uniqueness of the limit: just take $x_n = y_n$ for all n.)

● *Sandwich principle*

Suppose that for all $n \geq n_0$ (for some fixed n_0) we have $x_n \leq y_n \leq z_n$, and also that $x_n \to x$, $z_n \to x$. Then $y_n \to x$.

PROOF
Given $\varepsilon > 0$ we have to show that $|y_n - x| < \varepsilon$ for large enough n, i.e. that either $0 \leq y_n - x < \varepsilon$ or $0 \leq x - y_n < \varepsilon$. But for $n \geq n_0$ we have both $y_n - x \leq z_n - x$ and

$x - y_n \leq x - x_n$ and both these bounds can be made $< \varepsilon$ by choosing n large enough.

⊕ *Example 11*

Let $y_n = (100 + 5^n)^{1/n} = 5(\frac{100}{5^n} + 1)^{1/n}$, which lies between the constant sequence $x_n = 5 = 5(0 + 1)^{1/n}$ and $z_n = 5(20 + 1)^{1/n}$, which converges to 5. Therefore $\lim_{n \to \infty} y_n = 5$.

EXERCISES ON 2.2

1. Show that the following sequences diverge:
   (i) $x_n = (3n + (-1)^n)$ (ii) $x_n = \frac{(-1)^n n}{2+n}$
2. Suppose that (x_n) and (y_n) are null sequences. Give an example where the ratio $(\frac{x_n}{y_n})$ converges and one where it diverges.
3. Prove that $x_n = \frac{\sin n}{\sqrt{n}}$ is a null sequence.

Summary

We have reviewed the fundamental concept of the *limit* of a real sequence, applied it to finding the *sum* of an infinite series and described how the process of finding the limit interacts with the algebraic operations (addition, multiplication and their inverses) on real numbers and also with the order properties of the real line $\mathbb{R}$. In doing so we have highlighted the role played by *inequalities*, which enable us to compare the relative sizes of various expressions, and we have shown how a quite small stock of basic examples can be combined with our theorems to handle much more complicated sequences.

The results we have found so far would work equally well if we restricted ourselves to the set $\mathbb{Q}$ of rational numbers; we have not yet used the *Completeness Axiom* which distinguishes the ordered field $\mathbb{R}$ from $\mathbb{Q}$. That will be the focus of the next chapter.

FURTHER EXERCISES ·

1. Combining the results proved for null sequences, show that:
   (i) $(n^p x^n)$ is null whenever $|x| < 1$ and $p > 0$.
   (ii) $(\frac{x^n}{n!})$ is null for all real numbers x.
   (iii) $(\frac{n^p}{n!})$ is null for all $p > 0$.
2. Let

$$x_n = \frac{n! - (-1)^n}{3^n + 4(n!)}$$

Use the previous exercises to justify the claim that

$$x_n = \frac{\frac{1}{4} - \frac{(-1)^n}{4(n!)}}{\frac{3^n}{4(n!)} + 1} \to \frac{1}{4}$$

as $n \to \infty$.

3. Show that if $x_n \to x$ then $|x_n| \to |x|$. Is the converse true?

4. Find the sums of the following series by calculating their n^{th} partial sums:

(i) $\sum_{n \geq 1} \frac{1}{n(n+1)}$ (ii) $\sum_{n \geq 1} \frac{n}{(n+1)!}$

5. Test your skills against those of Nicole Oresme (1323–82) by explaining why it should be true that

$$\frac{1}{2} + \frac{2}{4} + \frac{3}{8} + \dots + \frac{n}{2^n} + \dots = 2$$

Oresme's idea was to represent the left-hand side by drawing a unit square, adding half a unit square on the rear half, then a quarter of a unit square on the end of that, etc. He then drew in vertical lines to describe the area of this 'infinite tower'. What happens when the tower collapses?

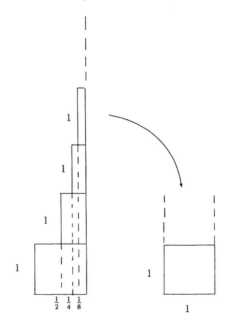

Fig. 2.1 Oresme's Tower

3 • Completeness and Convergence

In finding limits of sequences we have so far had to 'guess' the limit first, and then gone on to *prove* that our guess was correct. It would be much more satisfactory if we could find a *recipe* for deducing that certain types of sequences will always converge. By the same token it will be useful later to have a series of *tests* which can be used to decide when a given series converges – even if the tests will usually not be able to give us much information on the value of the sum of the series. In all of this the reason why the set $\mathbb{R}$ of real numbers is used as our number system, in preference to (e.g.) $\mathbb{Q}$, will become clearer, since all of the recipes and tests make essential use of the fact that $\mathbb{R}$ is *complete* and therefore contains no awkward 'gaps'.

3.1 Completeness and sequences

It is clearly too much to expect a general recipe which will always decide for us whether a given sequence converges. However, partial sums of series with positive terms provide a clue: the partial sums (s_n) of such series keep increasing if we add more and more terms, i.e. $s_n \leq s_{n+1}$ for all $n \in \mathbb{N}$. Thus the series 'should' either converge or 'go to infinity', i.e. if the sequence (s_n) is bounded above then it should converge. Before proving this claim we record some simple definitions:

• Definition 1

A real sequence $(x_n)_{n \geq 1}$ is called

(i) *increasing* if $x_n \leq x_{n+1}$ for every $n \geq 1$
(ii) *decreasing* if $x_n \geq x_{n+1}$ for every $n \geq 1$.

If either (i) or (ii) hold we say the sequence (x_n) is *monotone*. If the inequalities are strict the sequence is called *strictly* increasing (resp. decreasing, monotone).

If we restrict ourselves to rational numbers, then it is *not true* that every increasing bounded sequence converges: the increasing sequence 1.5, 1.75, 1.732, ... of rational numbers gets ever closer to $\sqrt{3}$, but never reaches a limit in $\mathbb{Q}$, since $\sqrt{3}$ does not belong to $\mathbb{Q}$. Nor should we attempt (as the famous nineteenth century French mathematician Augustin-Louis Cauchy once did!) to 'define' $\sqrt{3}$ as the limit of the approximating sequence: our definition of the limit of a real sequence (Definition 1 in Chapter 2) *presupposes* that we can find this limit *within* the number system in which we work.

It is to plug these gaps in the 'number line' that we have to go beyond the set $\mathbb{Q}$ for our number system, to define the set $\mathbb{R}$ of *real numbers*. We shall not give a full definition or construction of $\mathbb{R}$ (*see* [*NSS*], Chapter 5); we shall take knowledge of the algebraic and order structure of $\mathbb{R}$ for granted. But to fill in the gaps we need to restate the axiom which distinguishes $\mathbb{R}$ from $\mathbb{Q}$:

● Completeness Axiom

Let A be a non-empty subset of $\mathbb{R}$. If A is bounded above, then it has a least upper bound in $\mathbb{R}$.

This requires some explanation. A set $A \subset \mathbb{R}$ is *bounded above* if there exists $x \in \mathbb{R}$ such that $a \leq x$ for every $a \in A$; in that case x is called an *upper bound* for A. The least element z of the set of all upper bounds for A is then the *supremum* or *least upper bound* for A. In other words, we must have $a \leq z$ for all $a \in A$, and if x is any upper bound for A then $z \leq x$. We write this as: $z = \sup A$.

TUTORIAL PROBLEM 3.1

Discuss similar definitions for the following: the set $B \subset \mathbb{R}$ is bounded below; y is the *greatest lower bound* (also called the *infimum*) of B. We write this as: $y = \inf B$.

Show that $\inf B = -\sup\{-b : b \in B\}$.

It is now simple to prove the *Principle of Archimedes,* which we can restate as follows:

The set $\mathbb{N}$ is unbounded above in $\mathbb{R}$, i.e. given any real number x we can find $n \in \mathbb{N}$ such that $n > x$.

PROOF
$\mathbb{N}$ is a subset of $\mathbb{R}$; hence if it were bounded above it would have a least upper bound, M say. Hence there would be an $m \in \mathbb{N}$ with $m > M - 1$ (since $M - 1$ is less than M and so cannot be an upper bound for $\mathbb{N}$). But then $n = m + 1 > M$ and $n \in \mathbb{N}$, which contradicts M being an upper bound for $\mathbb{N}$. Hence $\mathbb{N}$ is unbounded above.

TUTORIAL PROBLEM 3.2

Read Sections 5.3 and 5.4 of [*NSS*]. In particular, convince yourself that the Completeness Axiom guarantees the *existence* of irrationals in R, and that between every two given rational numbers there are (infinitely many) rational and (infinitely many) irrational numbers.

We need some further notation to distinguish the subsets of $\mathbb{R}$ which inherit all the 'nice' properties of $\mathbb{R}$:

A *closed interval* in $\mathbb{R}$ is a set of the form

$$[a, b] = \{x \in \mathbb{R} : a \leq x \leq b\}$$

whereas an *open interval* in $\mathbb{R}$ has the form

$$(a, b) = \{x \in \mathbb{R} : a < x < b\}$$

The definitions of the *half-open* intervals $[a, b)$ and $(a, b]$ should now be obvious. In all the above we assume that $a < b$, to avoid trivialities (when $a = b$ only the 'closed interval' $[a, a] = \{a\}$ is a non-empty set). The result discussed in Tutorial Problem 3.2 can now be restated as:

> *Any interval in* $\mathbb{R}$ *contains both rational and irrational numbers* (and, in fact, contains infinitely many numbers of both kinds).

The recipe for finding the limit of a bounded increasing sequence can be stated concisely:

● *Theorem I* ─────────────

Let (x_n) be an increasing sequence of real numbers and suppose (x_n) is bounded above in $\mathbb{R}$. Then (x_n) converges to its supremum as $n \to \infty$.

PROOF

Since (x_n) is bounded above, its least upper bound $\alpha = \sup_n x_n$ exists as a real number. Suppose $\varepsilon > 0$ is given; since α is the *least* upper bound of (x_n) there must exist $N \in \mathbb{N}$ such that $x_N > \alpha - \varepsilon$. But as $x_n \leq x_{n+1}$ for all n, it follows that $\alpha \geq x_n \geq x_N > \alpha - \varepsilon$ whenever $n > N$. Thus $|x_n - \alpha| < \varepsilon$ for all $n > N$, hence (x_n) converges to α as $n \to \infty$.

It is now easy to use Tutorial Problem 3.1 and Theorem 1 to show that if (x_n) is decreasing and bounded below, then $\lim_{n \to \infty} x_n = \inf_n x_n$.

The two previous results together imply:

> *Every bounded monotone sequence in* $\mathbb{R}$ *is convergent.*

Our 'recipe' for finding the limit of a sequence (x_n) is therefore to show that (x_n) is: (i) monotone, (ii) bounded. When these conditions are satisfied, $\lim_n x_n$ equals $\sup_n x_n$ when (x_n) is increasing, and $\inf_n x_n$ when (x_n) is decreasing. So we need to consider ways of identifying which sequences are monotone.

⊕ *Example I*

To decide whether a sequence (x_n) is monotone, we need to compare x_n and x_{n+1} for every $n \in \mathbb{N}$. We can do this directly, by finding the sign of their difference $(x_{n+1} - x_n)$; alternatively, we shall often consider the size of the ratios $\frac{x_{n+1}}{x_n}$. If the ratios remain less than 1 from some n_0 onwards, it follows that $x_{n+1} < x_n$ for $n \geq n_0$. Thus the sequence $(x_n)_{n \geq n_0}$ is monotone decreasing. If we can also find a lower bound for (x_n) we know from the above that $\lim_{n \to \infty} x_n = \inf_n x_n$. Then it only remains to find the infimum, which is often somewhat easier.

For example, let $x_n = \frac{n}{2^n}$ for $n \geq 1$. We have $\frac{x_{n+1}}{x_n} = \frac{(n+1)}{2^{n+1}} \times \frac{2^n}{n} = \frac{1}{2}(\frac{n+1}{n})$, which is less than 1 whenever $n > 1$. In fact, the largest ratio is $\frac{x_2}{x_1} = 1$, so the sequence $(x_n)_{n \geq 1}$ is decreasing. Since each term is positive, the sequence is also bounded below by 0. Hence by the remark following Theorem 1 it must converge to $\inf_n x_n = x \geq 0$. To show that $x = 0$, consider the ratios again: we have

$x_{n+1} \leq \frac{1}{2}(\frac{n+1}{n})x_n$ for all $n \geq 1$, and both sides converge as $n \to \infty$. In fact, both $x_{n+1} \to x$ and $x_n \to x$ as $n \to \infty$, while $\frac{n+1}{n} = 1 + \frac{1}{n} \to 1$. Hence $0 \leq x \leq \frac{1}{2}x$, which means that $x = 0$.

Example 2

It is not much harder to prove that $x_n = \frac{n^k}{(1+\frac{1}{k})^n}$ is a null sequence for *any* fixed $k \in \mathbb{N}$. The technique is the same:

$$\frac{x_{n+1}}{x_n} = \frac{(n+1)^k}{(1+\frac{1}{k})^{n+1}} \times \frac{(1+\frac{1}{k})^n}{n^k} = \frac{(1+\frac{1}{n})^k}{(1+\frac{1}{k})} < 1$$

whenever N is large enough to ensure that for all $n \geq N$, $(1+\frac{1}{n})^k < 1 + \frac{1}{k}$. Such N can be found, since (by Theorem 1 in Chapter 2) $(1+\frac{1}{n})^k \to 1$ as $n \to \infty$. Hence the sequence $(x_n)_{n \geq N}$ is decreasing, bounded below by 0, and thus converges. The limit is again 0, by the same argument as in the previous example.

It is instructive, however, to calculate the value of N for larger values of k: even if $k = 100$, so that $x_n = \frac{n^{100}}{(1.01)^n}$, we need $N = 10\,050$, and the sequence reaches values of the order of 10^{365} before 'turning round and heading for zero'!

TUTORIAL PROBLEM 3.3

Write a simple computer program to illustrate graphically how this sequence changes for successive values of n. The Appendix contains a Pascal program which displays the behaviour of this and related null sequences.

Recursively defined sequences

Iteration, or *recursion*, as we shall call it, is what the computer does best. The discussion below provides a sound theoretical basis for numerical techniques which you may well have encountered earlier. A number of examples is given in [*NSS*], Chapter 7. Here we concentrate more on the underlying ideas.

Given a recursively defined sequence (x_n) with $x_{n+1} = f(x_n)$, it is often possible to show fairly simply that (x_n) is monotone and bounded. Hence it must converge to a limit $x = \lim_n x_n$, and this limit is frequently a *fixed point* of the function f which defines the recursion, i.e. we have $x = f(x)$, which in turn can enable us to find the value of x. The following basic examples illustrate the technique – note that in these cases it is usually simpler to use induction to compare x_n and x_{n+1} directly, rather than considering their ratio.

Example 3

Suppose that $x_1 = 1$, and for $n \geq 1$, $x_{n+1} = \sqrt{1 + x_n}$. Thus (x_n) begins with: $1, \sqrt{2},$ $\sqrt{1 + \sqrt{2}}, \sqrt{1 + \sqrt{1 + \sqrt{2}}}$, etc. It is certainly not immediately clear what the limit will be if it exists! However, the first few terms do suggest that (x_n) is increasing. We prove this in general, using induction: $x_1 = 1 < \sqrt{2} = x_2$. Now if $x_{n-1} < x_n$, then $x_n = \sqrt{1 + x_{n-1}} < \sqrt{1 + x_n} = x_{n+1}$, hence (x_n) is increasing, by induction. Also,

$x_1 < 2$, and if $x_n < 2$ then $x_{n+1} = \sqrt{1 + x_n} < \sqrt{1+2} = \sqrt{3} < 2$ also. So (x_n) is bounded above by 2.

We have shown that (x_n) is increasing and bounded above, hence $x = \lim_n x_n$ exists in $\mathbb{R}$. To find its value we first need to find $\lim_n \sqrt{1 + x_n}$ as $n \to \infty$. In fact, this limit is simply $\sqrt{1 + x}$. To prove this, write

$$|\sqrt{1 + x_n} - \sqrt{1 + x}| = \frac{|1 + x_n - (1 + x)|}{\sqrt{1 + x_n} + \sqrt{1 + x}} < |x_n - x|$$

since x_n and x are positive, so the denominator is greater than 1. But $x_n \to x$, hence $\sqrt{1 + x_n} \to \sqrt{1 + x}$ as claimed.

Now we have $x = \lim_n x_n = \lim_n x_{n+1} = \sqrt{1 + x}$, i.e. $x^2 - x - 1 = 0$, so that $x = \frac{1 + \sqrt{5}}{2}$ (since $x > 0$). This number is known as the *golden section*, and we will meet it again as the limit of ratios of successive Fibonacci numbers.

● *Example 4*

(Harder!) Consider the sequence given by $x_0 = a + 1$, $x_{n+1} = x_n(1 + \frac{a - x_n^k}{kx_n^k})$ for a fixed $a > 0$ and fixed $k \in \mathbb{N}$. Then: $x_n > 0$, $x_{n+1} < x_n$ and $x_n^k > a$. To see this, we again use induction. All three claims are true with $n = 0$, and we assume them for some fixed n. Now since $x_n > 0$, $a < x_n^k$ means that $x_{n+1} < x_n$; however x_{n+1} remains positive, since $kx_n^k + a - x_n^k > 0$ for all $k \geq 1$; and finally we can use the Bernoulli inequality with $x = \frac{a - x_n^k}{kx_n^k}$ (which is greater than -1 because $kx_n^k + a - x_n^k > 0$) to conclude that

$$x_{n+1}^k = x_n^k \left(1 + \frac{a - x_n^k}{kx_n^k}\right)^k > x_n^k \left\{1 + k\left(\frac{a - x_n^k}{kx_n^k}\right)\right\} = a$$

Hence by induction all three claims are true for all n. So (x_n) is decreasing and bounded below by 0. It therefore converges to $\alpha = \inf_n x_n$, which is non-negative. The recursion formula can be written: $kx_{n+1}x_n^{k-1} = (k - 1)x_n^k + a$, and by the algebra of limits we can let $n \to \infty$ on both sides to obtain $k\alpha\alpha^{k-1} = (k - 1)\alpha^k + a$. Hence $\alpha^k = a$, in other words $\alpha = a^{1/k}$. We have shown, therefore, that every positive real number has a positive k^{th} root for every fixed $k \in \mathbb{N}$. This root is *unique*, since if we have two positive numbers α, β with $\alpha < \beta$ then $0 < \alpha^k < \beta^k$, so α^k and β^k cannot both equal a.

How did we hit on this 'clever' recursion? In fact, it is just a version of *Newton's method* for finding approximate roots of equations. In particular, we can approximate roots of $f(x) = x^k - a$ by taking an overestimate x_n of this root, and choosing $x_{n+1} = x_n - \frac{f(x_n)}{f'(x_n)}$ as the next estimate (where f' denotes the derivative). In the case $f(x) = x^k - a$ we have $f'(x) = kx^{k-1}$, and we have just verified that this procedure *does* yield a convergent sequence whose limit is the positive root of the equation $x^k = a$.

TUTORIAL PROBLEM 3.4

Find out about Newton's method, and try to write a computer program to implement it. Now test your program on several examples.

Subsequences

The above examples should convince you of the importance of monotone sequences. But by no means all sequences are monotone, of course! For example, $-1, 1 - 1, 1 - 1, 1$ ... jumps 'up' and 'down' alternately, and has no limit, while $x_n = \frac{(-1)^n}{n}$ defines another non-monotone sequence which *does* have a limit, namely 0. In both cases the *subsequence* of *even* terms is monotone however: in the first case it is just the constant sequence $1,1,1,1, \ldots$, in the second it is the decreasing sequence $\frac{1}{2}, \frac{1}{4}, \frac{1}{6}, \ldots$

The meaning of 'subsequence' should be clear from the above: we 'pick out' some of the elements of the sequence (x_n), usually according to some fixed rule. For example, writing $n_r = r^2$ we can pick out the subsequence $(x_{n_r}) = \{x_1, x_4, x_9, x_{16}, \ldots\}$ from a sequence (x_n). In general, all we need is a *strictly increasing function* $f(r) = n_r$ so that we can ensure that $n_r \to \infty$ when $r \to \infty$.

◉ *Example 5*

A sequence can have many interesting subsequences: consider the following, which simply lists the rationals between 0 and 1 in increasing order of their denominators, with increasing numerators within each group:

$$(x_n) = \left\{ \frac{1}{2}, \frac{1}{3}, \frac{2}{3}, \frac{1}{4}, \frac{2}{4}, \frac{3}{4}, \frac{1}{5}, \frac{2}{5}, \frac{3}{5}, \frac{4}{5}, \frac{1}{6}, \frac{2}{6}, \ldots \right\}$$

Among its subsequences we find the following:

(i) $\frac{1}{2}, \frac{1}{3}, \frac{1}{4}, \ldots$ which decreases to 0;
(ii) $\frac{1}{2}, \frac{2}{3}, \frac{3}{4}, \ldots$ which increases to 1;
(iii) $\frac{1}{2}, \frac{2}{4}, \frac{3}{6}, \ldots$ which is the constant sequence $\frac{1}{2}$;
(iv) for each real number x between 0 and 1 there is a subsequence of (x_n) which converges to x. When $x \in \mathbb{Q}$ this is easy to see, since we need only choose $x, \frac{2x}{2}, \frac{3x}{3}, \ldots$ When y is irrational we can choose a sequence (y_n) of rationals converging to y, and for each n choose the subsequence $y, \frac{2y}{2}, \frac{3y}{3}, \ldots$ as above. The sequence composed of all these subsequences, taken one after another, then converges to y.

As you should by now have guessed, all subsequences of convergent sequences will converge to the same limit:

● *Proposition I*

Suppose that $x_n \to x$. Then every subsequence (x_{n_r}) of (x_n) also converges to x.

PROOF
Given $\varepsilon > 0$, find $N \in \mathbb{N}$ such that $|x_n - x| < \varepsilon$ for $n > N$. Since (n_r) is *strictly* increasing, we can find R so large that $n_r > N$ whenever $r > R$. Hence $|x_{n_r} - x| < \varepsilon$ whenever $r > R$, so $x_{n_r} \to x$ as $r \to \infty$.

In particular, this shows that we need only find two subsequences with *different* limits to prove that a sequence *fails* to converge: for example, (x_n) with $x_n = (-1)^n$ has subsequences (x_{2n}) and (x_{2n-1}) converging to 1 and -1 respectively; thus (x_n) diverges.

Example 6

A more subtle example is the following: let $x_n = \sin(n)$ for all $n \in \mathbb{N}$. Now $\sin(x) \geq \frac{1}{2}$ when $\frac{\pi}{6} < x < \frac{\pi}{2}$, and $\sin(x) \leq -\frac{1}{2}$ when $\frac{7\pi}{6} < x < \frac{11\pi}{6}$, so $n_1 = 1$, $m_1 = 4$ provides natural numbers such that $\sin(n_1) > \frac{1}{2}$ and $\sin(m_1) < -\frac{1}{2}$ respectively. Since $\sin(x + 2\pi) = \sin(x)$ for all x, we can move along in intervals of length 2π to find sequences (n_r) and (m_r) of natural numbers with $\sin(n_r) > \frac{1}{2}$ and $\sin(m_r) < -\frac{1}{2}$ for all $r \geq 1$. On the other hand, if $x = \lim_n x_n$ existed, we would also have $x = \lim_r x_{n_r} = \lim_r x_{m_r}$ and we would need simultaneously that $x \geq \frac{1}{2}$ and $x \leq -\frac{1}{2}$, which is impossible. Therefore (x_n) diverges.

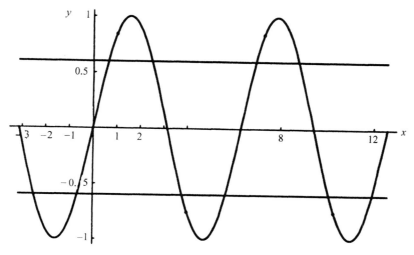

Fig. 3.1. Subsequences from sin x

The *existence* of monotone subsequences is surprisingly easy to prove; *finding* them in particular cases is not always that easy, though!

• Proposition 2

Every real sequence has a monotone subsequence.

PROOF

Suppose a sequence (x_n) is given. To pick out 'large' elements, let $C = \{N \in \mathbb{N} : x_m < x_N \text{ for all } m > N\}$. This set is either bounded or unbounded above. We consider both cases separately.

If C is bounded above, then it contains only *finitely* many elements of $\mathbb{N}$. If C is empty, let $n_1 = 1$. Otherwise we can find $n_1 \in \mathbb{N}$ greater than the maximum of C, so *none* of the numbers $n_1, n_1 + 1, n_1 + 2, \ldots$ belong to C. But if n_1 is not in C there must exist $n_2 \in \mathbb{N}$ such that $n_2 > n_1$ and $x_{n_2} \geq x_{n_1}$. Now n_2 is greater than n_1, so it cannot belong to C either. So we can find $n_3 > n_2$ for which $x_{n_3} \geq x_{n_2}$. Hence the subsequence $\{x_{n_r} : r = 1, 2, 3, \ldots\}$ constructed in this way is *increasing*. On the other hand, if C is unbounded above, we can find an *infinite* sequence of natural numbers $n_1 < n_2 < n_3 < \ldots$ all of which lie in C. But by definition of C, $n_2 > n_1$

means that $x_{n_2} \leq x_{n_1}$, $n_3 > n_2$ implies $x_{n_3} \leq x_{n_2}$, etc. So the subsequence found in this way is *decreasing*.

● Theorem 2—Bolzano–Weierstrass Theorem ─────

Every bounded monotone sequence in $\mathbb{R}$ has a convergent subsequence.

PROOF
Given the sequence (x_n) choose a monotone subsequence (x_{n_r}). Since $\{x_n : n \in \mathbb{N}\}$ is bounded, so is its subset $\{x_{n_r} : r \in \mathbb{N}\}$. Hence (x_{n_r}) is a bounded monotone sequence, and therefore converges, by the remark following Theorem 1.

This seemingly innocent result will have far-reaching consequences when we consider more general subsets of $\mathbb{R}$ and functions $\mathbb{R} \mapsto \mathbb{R}$. Our first application verifies more generally something we have already observed when dealing with 'oscillating' sequences such as (x_n) with $x_n = (-1)^n$ for each $n \geq 1$.

● Proposition 3

In every bounded divergent sequence (a_n) we can find two subsequences which converge to different limits.

(This provides a partial converse to Proposition 1.)

PROOF
Given a bounded sequence (a_n) find a convergent subsequence (a_{n_r}) by the Bolzano–Weierstrass Theorem. Let $a = \lim_{r \to \infty} a_{n_r}$. As (a_n) is divergent, it cannot converge to a. Hence there exists $\varepsilon > 0$ such that for all $N \in \mathbb{N}$ we have $|a_n - a| > \varepsilon$ for at least some $n > N$. Apply this successively with $N = 1, 2, 3, \ldots$ to obtain a new subsequence (a_{m_r}) with $|a_{m_r} - a| > \varepsilon$ for every $r \geq 1$. Then (a_{m_r}) is a bounded sequence, hence by the Bolzano–Weierstrass Theorem it has a convergent subsequence, (b_k) say. The limit of (b_k) cannot be a, since $(b_k) \subset (a_{m_r})$ and these are all at least ε away from a. Thus (a_{n_r}) and (b_k) are convergent subsequences of (a_n) with different limits.

Thus, among bounded sequences, convergent sequences are *exactly* those for which all subsequences have the same limit. To deal with the variety of possible limits of subsequences of divergent sequences we introduce a new concept and some notation:

● Definition 2

(i) A real number a is an *accumulation point* of the sequence (x_n) if some subsequence of (x_n) converges to a.

(ii) Suppose (x_n) is a given real sequence. If (x_n) converges, then $x = \lim_{n \to \infty} x_n$ is its *only* accumulation point, since for each $\varepsilon > 0$ there are at most finitely many n for which $|x_n - x| \geq \varepsilon$. We shall write $N(x, \varepsilon)$ for the open interval $(x - \varepsilon, x + \varepsilon)$ of radius ε centred at x (we call this the *ε-neighbourhood* of x), and say that *nearly all* elements of (x_n) belong to $N(x, \varepsilon)$ if the above statement is true. Thus $x = \lim_{n \to \infty} x_n$ is rephrased more loosely as the statement: for all $\varepsilon > 0$ nearly all $x_n \in N(x, \varepsilon)$.

The condition that a is an accumulation point of (x_n) is much weaker: all we require is that for each $\varepsilon > 0$ there are *infinitely many* n_r for which $|x_{n_r} - a| < \varepsilon$, since this will ensure that the subsequence $(x_{n_r})_{r \geq 1}$ converges to a. In particular, given any $\varepsilon > 0$, the neighbourhood $N(a, \varepsilon) = (a - \varepsilon, a + \varepsilon)$ will then contain infinitely many elements of the sequence (x_n) – but not necessarily nearly all x_n, since any divergent sequence will have more than one accumulation point: the sequence of Example 5 has *every* $x \in [0, 1]$ as an accumulation point! (This shows that events that happen 'infinitely often' will nonetheless sometimes happen only 'rarely' among the elements of a sequence: the subsequence $\{\frac{1}{2}, \frac{2}{4}, \frac{3}{6}, \ldots, \frac{n}{2n}, \ldots\}$ becomes pretty sparsely distributed within the full sequence of Example 5.)

It is convenient to allow unbounded sequences also, at the cost of writing $x_n \to \infty$ or $x_n \to -\infty$, where this notation is interpreted as indicated after Definition 1 in Chapter 2. We then allow $\pm\infty$ as accumulation points of (x_n). With this extended interpretation of 'convergence' the Bolzano–Weierstrass Theorem can be restated as:

> *Every real sequence has a convergent subsequence, and therefore has at least one accumulation point.*

The extension is clear: if (x_n) is unbounded above (resp. below) we can find a subsequence $x_{n_r} \to \infty$ (resp. $x_{n_r} \to -\infty$) as $r \to \infty$.

TUTORIAL PROBLEM 3.5

Decide what is meant by a 'neighbourhood of ∞' if we apply the same conventions to the notation $x_n \to \infty$.

• *Definition 3*

Let (x_n) be a given real sequence. Define the *upper limit* $\beta = \overline{\lim}_n x_n$ by setting $\beta = \inf\{b : x_n \leq b$ for nearly all $n \in \mathbb{N}\}$. (β is often also denoted by $\limsup_n x_n$.)

It is not hard to see that β is the *largest* accumulation point of the sequence (x_n): clearly, for any $\varepsilon > 0$ we cannot construct a subsequence converging to $\gamma = \beta + \varepsilon$, since for all except finitely many n, $x_n \leq \beta = \gamma - \varepsilon$. Hence it suffices to show that β is an accumulation point. But by definition $x_n > \beta - \varepsilon$ must hold for infinitely many n, while nearly all $x_n \leq \beta + \varepsilon$, hence infinitely many $x_n \in N(\beta, \varepsilon)$, which completes the proof.

EXERCISES ON 3.1

1. Let $x_1 = 2.5$, $x_{n+1} = \frac{1}{5}(x_n^2 + 6)$ for $n \geq 1$. Show that for $n \geq 1$

$$x_{n+1} - x_n = \frac{1}{5}(x_n - 2)(x_n - 3)$$

   Show by induction that each element of the sequence lies between 2 and 3. Is this sequence monotone? Find its limit as $n \to \infty$.
2. We have shown that a bounded sequence (x_n) either converges or has at least two subsequences which converge to different limits. In this latter case we say

that (x_n) *oscillates boundedly*. Investigate the various possibilities for unbounded sequences by considering the following examples:

(i) $x_n = (-n)^n$

(ii) $x_n = 2^n$

(iii) $x_n = -2^n$

(iv) $x_n = \begin{cases} n^n & n \text{ even} \\ 0 & n \text{ odd} \end{cases}$

3. (i) Define the *lower limit* $\alpha = \sup\{a : x_n \geq a \text{ for nearly all } n\}$. (We write $\alpha = \underline{\lim}_n x_n$.) Show that α is the *smallest* accumulation point of (x_n).

(ii) Show that for any given $\varepsilon > 0$,

$$\underline{\lim}_n x_n - \varepsilon < x_n < \overline{\lim}_n x_n + \varepsilon$$

for nearly all $n \in \mathbb{N}$.

(iii) Deduce that (x_n) converges if and only if $\underline{\lim}_n x_n = \overline{\lim}_n x_n$. (Here we allow $\pm\infty$ as limits, by convention.)

4. Find the upper and lower limits of the following sequences (x_n):

(i) $x_n = (-1)^n(1 + \frac{1}{n})$

(ii) The sequence of Example 5.

Compare your answers with $\sup_n x_n$ and $\inf x_n$ in each case.

5. Let $a_1 = 1$, $a_{n+1} = \frac{1}{2+a_n}$ for $n \geq 1$. Show that $a_n \in [0, 1]$ for all n.

Set $b_n = a_{2n-1}$ and $c_n = a_{2n}$ for each n. Show that the differences $(a_{n+2} - a_n)$ and $(a_{n+1} - a_{n-1})$ always have opposite signs and hence that (b_n) decreases while (c_n) increases. Expressing each of b_n and c_n in terms of the other, show that (a_n) converges and find its limit.

3.2 Completeness and series

Tests for convergence: Comparison, Ratio and Root tests

We have already used the fact that the partial sums of series with positive terms form an increasing sequence; if $a_n > 0$ then obviously $s_{n+1} > s_n$, where $s_n = \sum_{i=1}^{n} a_i$. This provides a useful of *test of convergence* for such series.

● *Theorem 3—Comparison Test for series* ────────────

Suppose that $0 \leq a_i \leq b_i$ for all $i \in \mathbb{N}$. Then:

(i) if $\sum_i b_i$ converges, so does $\sum_i a_i$, and $\sum_{i=1}^{\infty} a_i \leq \sum_{i=1}^{\infty} b_i$

(ii) if $\sum_i a_i$ diverges, so does $\sum_i b_i$.

PROOF

Let $s_n = \sum_{i=1}^{n} a_i$ and $t_n = \sum_{i=1}^{n} b_i$ for each $n \in \mathbb{N}$. Then $0 \leq s_n \leq t_n$ for each n, hence if (t_n) is bounded above, so is (s_n). Hence $s_n \to \sup_n s_n \leq \sup_n t_n$. On the other hand, if $\sum_i a_i$ diverges, (s_n) is unbounded above, hence so is (t_n), and thus $\sum_i b_i$ diverges.

● *Example 7*

(i) $0 < \frac{1}{n^2} \leq \frac{1}{(n-1)n}$ for all $n > 1$. Now the terms on the right have sum $\sum_{n=2}^{\infty} \frac{1}{(n-1)n} = \sum_{m=1}^{\infty} \frac{1}{m(m+1)} = 1$, by Further Exercise 4(i) on Chapter 2. This implies that $\sum_n \frac{1}{n^2}$ converges, by comparison, and that $\sum_{n=2}^{\infty} \frac{1}{n^2} \leq 1$, i.e. $1 + \frac{1}{4} + \frac{1}{9} + \frac{1}{16} + \ldots = \sum_{n=1}^{\infty} \frac{1}{n^2} \leq 2$.

What the *exact* sum is here is rather more difficult to establish. In Chapter 4 we shall see how the eighteenth century mathematician Leonhard Euler arrived at the right answer, though his methods would not pass muster today – the purpose of this book, after all, is to instil a sense of rigour and precision in dealing with the infinite!

Note that in this example we compared series term by term from $n = 2$, rather than from $n = 1$. It should be clear by now that the Comparison Test applies equally to series with $a_n \leq b_n$ for $n \geq n_0$ for some n_0 – though we need to remember that this will only yield $\sum_{n=n_0}^{\infty} a_n \leq \sum_{n=n_0}^{\infty} b_n$, and that the first n_0 terms have to be handled separately in that case.

(ii) $\sum_n \frac{n^2+5n-1}{n^4+3}$ has $\frac{n^2+5n-1}{n^4+3} \leq \frac{6n^2}{n^4} = \frac{6}{n^2}$, so the series will converge by comparison with $\sum_n \frac{1}{n^2}$. On the other hand, $\sum_n \frac{n^2-5n+2}{n^3+3n^2+1}$ will diverge, since $\frac{n^2-5n+2}{n^3+3n^2+1} \geq \frac{\frac{1}{2}n^2}{n^3}$ for $n > 10$ (we have $n^2 - 5n = n(n-5) > \frac{1}{2}n^2$ if $n > 10$), and the series dominates the harmonic series term by term.

(iii) $\sum_n (\frac{1}{n} + \frac{1}{2})^n$ satisfies $(\frac{1}{n} + \frac{1}{2}) \leq \frac{1}{3} + \frac{1}{2} = \frac{5}{6}$ for all $n \geq 3$, and therefore $(\frac{1}{n} + \frac{1}{2})^n \leq (\frac{5}{6})^n$ for all such n. Hence this series converges by comparison with the geometric series for $x = \frac{5}{6}$, and since $\sum_{n=3}^{\infty}(\frac{5}{6})^n = (\frac{5}{6})^3(\frac{1}{1-\frac{5}{6}}) = \frac{125}{36}$, it follows that $\sum_{n=1}^{\infty}(\frac{1}{n} + \frac{1}{2})^n \leq (1 + \frac{1}{2}) + (\frac{1}{2} + \frac{1}{2})^2 + \frac{125}{36} = \frac{215}{36} < 6$.

The next three tests are simple consequences of the Comparison Test, and are proved by describing conditions under which we can compare the series to a geometric one. We shall keep to series with positive terms for the time being. The test which is easiest to use – though not the most powerful one we shall discuss – is the following:

• *Theorem 4—Ratio Test*

Let $\sum_n a_n$ be a series with positive terms.

(i) If there exists $r < 1$ and $n_0 \in \mathbb{N}$ such that $\frac{a_{n+1}}{a_n} \leq 1$ for all $n > n_0$ then $\sum_n a_n$ converges.

(ii) If $\frac{a_{n+1}}{a_n} \geq 1$ for all n then $\sum_n a_n$ diverges.

Often more useful in practice is the following:

Limit Form of the Ratio Test

Let $\sum_n a_n$ be a series with positive terms for which $\alpha = \lim_n \frac{a_{n+1}}{a_n}$ exists in $\mathbb{R}$. Then we have:

(i) if $\alpha < 1$ then $\sum_n a_n$ converges;

(ii) if $\alpha > 1$ then $\sum_n a_n$ diverges;

(iii) if $\alpha = 1$ then the test is inconclusive.

PROOF

We prove the Theorem first. Since $\sum_{n=1}^{\infty} a_n$ is finite if and only if $\sum_{n=n_0+1}^{\infty} a_n$ is finite, we can assume that $\frac{a_{n+1}}{a_n} \leq r$ for all $n \geq 1$. Thus for each $n \geq 1$, $a_{n+1} \leq r a_n \leq r^2 a_{n-1} \leq \ldots \leq r^n a_1$. Therefore $\sum_n a_n$ converges by comparison with the geometric series $a_1 \sum_n r^n$, since $0 \leq r < 1$. On the other hand, if $\frac{a_{n+1}}{a_n} \geq 1$ then

$a_{n+1} \geq a_n$ and therefore (a_n) increases (from n_0 onwards) and hence cannot converge to 0, since $a_n > 0$.

The *Limit Form* now follows at once: if $\alpha < 1$ then $r = \frac{\alpha+1}{2} < 1$ and at least from some n_0 onwards we must have $|\frac{a_{n+1}}{a_n} - \alpha| < \frac{1-\alpha}{2}$, and therefore $\frac{a_{n+1}}{a_n} \leq r$, so that by (i) the series converges. On the other hand, if $\alpha > 1$ we can also choose n_0 so that for all $n > n_0$ we have $|\frac{a_{n+1}}{a_n} - \alpha| < \frac{\alpha-1}{2}$, so that $\frac{a_{n+1}}{a_n} > 1$ for all such n, hence the series diverges by (ii).

To show that 'anything can happen' when $\alpha = 1$, consider the series $\sum_n \frac{1}{n}$ and $\sum_n \frac{1}{n^2}$: in the first case we have $\frac{a_{n+1}}{a_n} = \frac{\frac{1}{n+1}}{\frac{1}{n}} = \frac{n}{n+1} = 1 - \frac{1}{n+1} \to 1$ as $n \to \infty$, and in the second we have $\frac{a_{n+1}}{a_n} = \frac{\frac{1}{(n+1)^2}}{\frac{1}{n^2}} = (\frac{n}{n+1})^2 = (1 - \frac{1}{n+1})^2 \to 1$ also. But the first series diverges while the second converges. So the test is inconclusive when $\alpha = 1$.

Before stating the next test, recall that we say that a sequence has a certain property *for infinitely many n* if there is an infinite subsequence for which each term has that property.

● *Theorem 5—Cauchy's* n^{th} *Root Test*

Let $\sum_n a_n$ be a series with positive terms. If there are $r < 1$ and $n_0 \in \mathbb{N}$ such that $\sqrt[n]{a_n} \leq r$ for all $n > n_0$, then $\sum_n a_n$ converges. If $\sqrt[n]{a_n} > 1$ for infinitely many n, then $\sum_n a_n$ diverges.

Upper Limit Form of the Test

Let $\beta = \overline{\lim}_n \sqrt[n]{a_n}$. If $\beta < 1$ then the series $\sum_n a_n$ converges, and if $\beta > 1$ the series $\sum_n a_n$ diverges. The test gives no information if $\beta = 1$.

PROOF
We prove the Theorem as stated first. In the second case we have $a_n \geq 1$ for infinitely many n, so (a_n) cannot converge to 0, hence $\sum_n a_n$ diverges. In the first case we have $a_n \leq r^n$ for all $n > n_0$, hence the series converges by comparison with $\sum_n r^n$.

The proof of the Limit Form is almost identical; the case $\beta = 1$ can again be checked using the series $\sum_n \frac{1}{n}$ and $\sum_n \frac{1}{n^2}$. Details are left to the reader.

TUTORIAL PROBLEM 3.6

The nth Root Test is *stronger* than the *Ratio Test*:
Verify that for any sequence (x_n) of positive reals,

$$\underline{\lim}_n \frac{x_{n+1}}{x_n} \leq \underline{\lim}_n \sqrt[n]{x_n} \leq \overline{\lim}_n \sqrt[n]{x_n} \leq \overline{\lim}_n \frac{x_{n+1}}{x_n}$$

(*Hint:* to prove the final inequality, set $\beta = \overline{\lim}_n \frac{x_{n+1}}{x_n}$, which we can suppose finite. Let $\gamma > \beta$, and find N such that $\frac{x_{n+1}}{x_n} \leq \gamma$ for $n \geq N$. Now show that for all $n > N$,

$$x_n \leq x_N \gamma^{n-N}, \quad i.e. \quad \sqrt[n]{x_n} \leq (\sqrt[n]{x_N \gamma^{-N}}).\gamma$$

and hence that $\overline{\lim}_n \sqrt[n]{x_n} \leq \gamma$ for all $\gamma > \beta$. The other case is similar.)

As we have seen, neither the Ratio Test nor the Root Test can distinguish between the convergent series $\sum_n \frac{1}{n^2}$ and the divergent series $\sum_n \frac{1}{n}$. Our next result will allow us finally to determine which powers of reciprocals give convergent series – this time we compare with the geometric series $\sum_n \frac{1}{2^n}$:

● *Proposition 4—Cauchy's Condensation Test*

If (a_n) is a non-negative decreasing sequence then $\sum_n a_n$ converges if and only if $\sum_n 2^n a_{2^n}$ converges.

PROOF

Since (a_n) is decreasing, it follows that if $2^k \leq n \leq 2^{k+1}$ then $a_{2^{k+1}} \leq a_n \leq a_{2^k}$. So for each fixed $k \in \mathbb{N}$ we have:

$$a_{2^k} + a_{2^k+1} + a_{2^{k+1}+2} + \ldots + a_{2^{k+1}-1} \leq 2^k a_{2^k}$$

and similarly

$$a_{2^k+1} + a_{2^k+2} + a_{2^{k+1}+3} + \ldots + a_{2^{k+1}} \geq 2^k a_{2^{k+1}} = \frac{1}{2}(2^{k+1} a_{2^{k+1}})$$

so that, by grouping the terms of the series $\sum_n a_n$ into 2^k terms for each $k \geq 1$, we see that $\sum_n a_n$ will converge by comparison with $\sum_k 2^k a_{2^k}$ whenever the latter converges, and conversely, $\sum_k 2^k a_{2^k}$ converges by comparison with $2 \sum_n a_n$ whenever that series converges.

● *Example 8*

Now we can consider $\sum_n \frac{1}{n^r}$ for each rational power r. (In fact, the results hold for irrational powers also, but first we'll have to decide what $x^{\sqrt{2}}$, etc. can possibly *mean*! So we stick to rational powers for now.) Clearly, if $r \leq 0$ the terms $\frac{1}{n^r}$ do not converge to 0, so the series must diverge. For $r > 0$ we can use the Condensation Test, which means we need to look at the geometric series $\sum_k 2^k (\frac{1}{2^{kr}}) = \sum_k (2^{1-r})^k$. This series converges if and only if $2^{1-r} < 1$, and this occurs exactly if $r > 1$. Hence we have proved the series $\sum_n \frac{1}{n^r}$ converges if and only if $r > 1$.

Note how this distinguishes between the two special cases ($r = 1$ or 2) which we discussed earlier.

● *Example 9*

Of the above tests, the Ratio Test is much the easiest one to use. It is particularly useful when there is much cancellation in the ratios $\frac{a_{n+1}}{a_n}$, as will be the case with powers and factorials. For example, consider the series defining the number e, namely $\sum_{n\geq 0} \frac{1}{n!}$: here we have $\frac{a_{n+1}}{a_n} = \frac{n!}{(n+1)!} = \frac{1}{n+1}$, which is $\leq \frac{1}{2}$ for $n \geq 1$. Hence this series converges by the Ratio Test. We can do rather more: for each fixed $x \in \mathbb{R}$ the series $\sum_n \frac{|x|^n}{n!}$ will converge, since for this series $\frac{a_{n+1}}{a_n} = \frac{n!}{(n+1)!} |x| = \frac{|x|}{n+1}$, which will be less than (say) $\frac{1}{2}$ for large enough n (take $n \geq 2|x|$, using the Archimedes Principle). Hence this series converges also. This *suggests* that the series $\sum_n \frac{x^n}{n!}$ will converge for each $x \in \mathbb{R}$, so that for each x we can take the *sum* of the series as the *value* of a certain function of x. But we need to wait a little longer to prove this plausible fact.

TUTORIAL PROBLEM 3.7

We saw in Further Exercise 5 on Chapter 2 that it has been known since the fourteenth century that $\sum_n \frac{n}{2^n}$ converges to 2. What about $\sum_n \frac{n^2}{2^n}$? Here we have $\frac{a_{n+1}}{a_n} = \frac{(n+1)^2}{2^{n+1}} \times \frac{2^n}{n^2} = \frac{1}{2}(1 + \frac{1}{n})^2 \to \frac{1}{2}$, hence the series converges. On the other hand $\sum_n \frac{n^n}{n!}$ diverges, since $\frac{a_{n+1}}{a_n} = \frac{(n+1)^{n+1}}{(n+1)!} \times \frac{n!}{n^n} = (1 + \frac{1}{n})^n > 1$ for all n. This raises the question: does $\lim_n (1 + \frac{1}{n})^n$ exist, and, if so, what is it? In fact, this limit is e. To prove this, we would need to verify the interesting identity:

$$\lim_n \sum_{k=0}^{n} \frac{1}{k!} = \lim_n \left(1 + \frac{1}{n}\right)^n$$

This is left as a challenge to the brave: you have all the tools you need for the job, but it needs quite a bit of care and patience.

EXERCISES ON 3.2

1. In each of the following cases, use appropriate tests to decide whether the series converges or diverges:
   (i) $\sum_{n \geq 1} \frac{n!}{n^n}$ (ii) $\sum_{n \geq 1} \frac{n^n}{n!}$ (iii) $\sum_{n \geq 1} \frac{n!(n+4)!}{(2n)!}$
2. Use comparisons with known series to investigate the convergence of
   (i) $\sum_{n \geq 6} (\frac{1}{n-5})^{\frac{6}{5}}$ (ii) $\sum_{n \geq 1} \frac{1 + \cos n}{2^n}$ (iii) $\sum_{n \geq 1} \frac{1}{2^n}(2^n + 3^n)^{\frac{1}{n}}$.

3.3 Alternating series

Much more care is needed when we deal with series whose terms are not necessarily positive: for example, is there a difference in the behaviour of

$$1 + \frac{1}{2} + \frac{1}{3} + + \frac{1}{n} + \dots$$

and

$$1 - \frac{1}{2} + \frac{1}{3} - \frac{1}{4} + \dots + (-1)^{n-1}\frac{1}{n} + \dots ?$$

The former series diverges, but might there be 'enough cancellation' between the terms of the latter to allow it to converge? Also: in the former case it seems plausible that the *order* in which we add terms together should not matter. In the second case it seems that this could go badly wrong: consider 'bracketing' terms in the order

$$1 - \frac{1}{2} - \frac{1}{4} + \left(\frac{1}{3} - \frac{1}{6} - \frac{1}{8}\right) + \left(\frac{1}{5} - \frac{1}{10} - \frac{1}{12}\right) + \dots$$

which we can apparently write as

$$\left(1 - \frac{1}{2}\right) - \frac{1}{4} + \left(\frac{1}{3} - \frac{1}{6}\right) - \frac{1}{8} + \left(\frac{1}{5} - \frac{1}{10}\right) - \frac{1}{12} \dots$$

and which in turn 'reduces' to

$$\frac{1}{2} - \frac{1}{4} + \frac{1}{6} - \frac{1}{8} + \frac{1}{10} - \frac{1}{12} \cdots = \frac{1}{2}\left(1 - \frac{1}{2} + \frac{1}{3} - \frac{1}{4} + \frac{1}{5} - \frac{1}{6} \cdots\right)$$

Thus, simply by writing the terms in a different order, we seems to have 'reduced' the sum of the series to half its original size! So: either the sum doesn't make sense, or it is actually 0, or something 'illegal' has been done here.

Let us examine these three possibilities in turn.

First, we can show that the series $\sum_{n\geq 1}(-1)^{n-1}\frac{1}{n}$ *does converge*. This follows from the following general test for series with *alternating* signs (sometimes called *Leibniz's Test*):

• *Theorem 6—Alternating Series Test* ─────────

Suppose the sequence (a_n) decreases to 0. Then the series $\sum_{n\geq 1}(-1)^n a_n$ converges.

PROOF

The kth partial sum of the series is $s_k = \sum_{n=1}^{k}(-1)^n a_n$. For *even k*, say $k = 2m$, we have

$$s_{2m} = (a_1 - a_2) + (a_3 - a_4) + \ldots + (a_{2m-1} - a_{2m})$$

which shows that (s_{2m}) increases, as each bracket is non-negative. But we can also write the partial sums as

$$s_{2m} = a_1 - (a_2 - a_3) - (a_4 - a_5)\ldots - (a_{2m-2} - a_{2m-1}) - a_{2m}$$

and since each bracket is again non-negative, as is a_{2m}, we have shown that $s_{2m} \leq a_1$. (We are only summing *finitely many terms* here, so these bracketing operations are justified!) Thus the even partial sums (s_{2m}) form an increasing sequence which is bounded above by a_1, and therefore $s = \lim_{m\to\infty} s_{2m}$ exists in $\mathbb{R}$. But then

$$s_{2m+1} = s_{2m} + a_{2m+1} \to s$$

also, since $a_{2m+1} \to 0$ as $m \to \infty$. This suffices to show that $s = \lim_{n\to\infty} s_n$, since, if $\varepsilon > 0$ is given, we can find $N, N' \in \mathbb{N}$ such that $|s_{2m} - s| < \varepsilon$ and $|s_{2m+1} - s| < \varepsilon$ whenever $m > M = \max(N, N')$. So for all $k > K = 2M + 1$ we have $|s_k - s| < \varepsilon$.

Clearly this test applies to $\sum_n(-1)^{n-1}\frac{1}{n} = -\sum_n(-1)^n\frac{1}{n}$, since $(\frac{1}{n})$ decreases to 0. Similarly it shows that $\sum_n(-1)^n\frac{1}{\log n}$ converges. (*Remember* that we need to check that (a_n) *decreases to* 0.)

To decide what the *sum* $s = \sum_{n=1}^{\infty}(-1)^{n-1}\frac{1}{n}$ might be, we can observe first of all that writing the series as $(1 - \frac{1}{2}) + (\frac{1}{3} - \frac{1}{4}) + (\frac{1}{5} - \frac{1}{6}) + \ldots$ produces a series whose partial sums are just the subsequence (s_{2n}) of the sequence of partial sums (s_n) of the original series, and since (s_n) converges to s, so must (s_{2n}). On the other hand, (s_{2n}) increases, hence we must have $s \geq s_6 = \frac{1}{2} + \frac{1}{12} + \frac{1}{30} = \frac{37}{60}$. Thus s *cannot* be 0, which rules out the second of our three possibilities! (Incidentally, we can consider $1 - (\frac{1}{2} - \frac{1}{3}) - (\frac{1}{4} - \frac{1}{5}) - \ldots$ instead, and thus use the subsequence (s_{2n-1}) of (s_n), which is bounded below by $s_5 = 1 - \frac{1}{6} - \frac{1}{20} = \frac{47}{60}$. Therefore $\frac{37}{60} \leq s \leq \frac{47}{60}$.

Digression (which may be omitted at a first reading)

We have done enough to show that 'something illegal' was done when we rearranged the terms of the series $\sum_n (-1)^n \frac{1}{n}$ without 'due care and attention'. We'll see below how to overcome this difficulty. Meanwhile, however, we are left wondering *how* we can determine the sum of $\sum_n (-1)^n \frac{1}{n}$. To do this, we need yet another convergence test, which is also very useful in other cases (*see* [*NSS*], Chapter 8):

● *The Integral Test*

Let $f : [0, \infty) \mapsto [0, \infty)$ be a decreasing function (i.e. if $x < y$ then $f(x) \leq f(y)$) and write $a_n = f(n)$ and $b_n = \int_1^n f(x)\, dx$ for each $n \in \mathbb{N}$. Then the series $\sum_n a_n$ converges if and only if the sequence $(b_n)_{n \geq 1}$ converges.

We need to explain the meaning of b_n: it is one of the easier results in the theory of *Riemann integration* (which we present in Chapters 10 and 11, where the properties used below are also proved) that each monotone function *can* be integrated over bounded intervals like $[1, n]$. So at least b_n is well-defined. Moreover, the integral is itself monotone: for two functions f and g with $f \leq g$ we can show that $\int_a^b f \leq \int_a^b g$. So for any $m \geq 2$ we obtain

$$f(m) = \int_{m-1}^m f(m)\, dx \leq \int_{m-1}^m f(x)\, dx \leq \int_{m-1}^m f(m-1)\, dx = f(m-1)$$

since f is decreasing on the interval $[m-1, m]$. Summing from $m = 2$ to $m = n$:

$$\sum_{m=2}^n f(m) \leq \sum_{m=2}^n \int_{m-1}^m f(x)\, dx \leq \sum_{m=2}^n f(m-1)$$

But the sum of the integrals is just $\int_1^n f(x)\, dx = b_n$ (this is another fact about integrals which is proved in Chapter 10) and the sums at either end are partial sums of the series $\sum_n a_n$. Hence we have:

$$\sum_{m=2}^n a_m \leq b_n \leq \sum_{m=1}^{n-1} a_m$$

Finally, since $f \geq 0$ the integrals are all positive, so the (b_n) form an increasing sequence. If $\sum_n a_n$ converges, the partial sums are bounded above, hence so is (b_n). Thus (b_n) converges. Conversely, if the sequence (b_n) converges, then it is bounded above, hence the partial sums $\sum_{m=2}^n a_m$ are bounded above, so the series $\sum_n a_n$ converges. This 'proves' the claim in the Integral Test.

Applying the Integral Test to the function $f(x) = \frac{1}{x}$ (so that $a_m = \frac{1}{m}$ and $b_n = \int_1^n \frac{dx}{x}$) we can argue as in the string of inequalities above, and with k and $k+1$ in place of n, to find that

$$\sum_{m=2}^k \frac{1}{m} \leq \int_1^k \frac{dx}{x} \leq \sum_{m=1}^{k-1} \frac{1}{m}$$

$$\sum_{m=2}^{k+1} \frac{1}{m} \leq \int_1^{k+1} \frac{dx}{x} \leq \sum_{m=1}^k \frac{1}{m}$$

We shall show in Chapter 11 that for any n, $\log_e n = \int_1^n \frac{dx}{x}$. Thus the second string leads us to: $\log_e(k+1) \leq \sum_{m=1}^{k} \frac{1}{m}$, and by the first this can be written as $1 + \sum_{m=2}^{k} \frac{1}{m} \leq 1 + \log_e k$. So, since the log function is increasing, we have:

$$\log_e k \leq \log_e(k+1) \leq \sum_{m=1}^{k} \frac{1}{m} \leq 1 + \log_e k$$

Now write $\gamma_k = 1 + \frac{1}{2} + \frac{1}{3} + \ldots + \frac{1}{k} - \log_e k$ so that by subtracting $\log_e k$ in each of the above we have: $0 \leq \gamma_k \leq 1$ for each $k \in \mathbb{N}$. Moreover,

$$\gamma_k - \gamma_{k+1} = \left(\sum_{m=1}^{k} \frac{1}{m} - \log_e k \right) - \left(\sum_{m=1}^{k+1} \frac{1}{m} - \log_e(k+1) \right)$$

which equals

$$\log_e(k+1) - \log_e k - \frac{1}{k+1} = \int_k^{k+1} \left(x - \frac{1}{k+1} \right) dx \geq 0$$

(The final step uses yet another property of the integral!)

So we have shown that (γ_n) is decreasing and bounded below by 0, and hence converges (the limit γ turns out to be 0.577 to three decimal places, and is known as *Euler's constant*). This, finally, helps us to calculate $\sum_{n=1}^{\infty}(-1)^{n-1}\frac{1}{n}$:

$$\sum_{n=1}^{2k}(-1)^{n-1}\frac{1}{n} = 1 - \frac{1}{2} + \frac{1}{3} \cdots - \frac{1}{2k}$$

$$= \left(1 + \frac{1}{2} + \frac{1}{3} + \frac{1}{4} + \ldots + \frac{1}{2k-1} + \frac{1}{2k} \right) - \left(1 + \frac{1}{2} + \frac{1}{3} + \frac{1}{4} \ldots + \frac{1}{k} \right)$$

$$= (\gamma_{2k} + \log_e 2k) - (\gamma_k + \log_e k) = \log_e 2 + \gamma_{2k} - \gamma_k$$

But since $\gamma_k \to \gamma$ and so $\gamma_{2k} \to \gamma$, it follows that

$$\lim_k \sum_{n=1}^{2k}(-1)^{n-1}\frac{1}{n} = \log_e 2$$

EXERCISES ON 3.3

1. Apply the Integral Test to show that $\sum_{n \geq 1} \frac{1}{n^r}$ converges if and only if $r > 1$. (*Hint:* integrate $f(x) = \frac{1}{x^r}$.)
2. Use Euler's constant to find an estimate for the first n such that n^{th} partial sum of the harmonic series is greater than (i) 10, (ii) 100, (iii) 1000. (Now do you pity the poor snail in Tutorial Problem 2.2?)

3.4 Absolute and conditional convergence of series

The problem with $\sum_n(-1)^n\frac{1}{n}$ arose from our attempt to *rearrange its terms* without proper care about the effect this might have on the sum. This is a problem which cannot arise in summing *finitely* many terms, and we 'feel' that it should not arise when all the terms of an infinite series are positive. How, then, can we characterize series whose terms *can* be rearranged without affecting the sum?

Again the above series (with $a_n = (-1)^{n-1}\frac{1}{n}$) provides a clue: in this case we know that $\sum_n a_n$ converges but $\sum_n |a_n|$ (i.e. the harmonic series) *diverges*. This leads to the following:

● Definition 4

A series $\sum_n a_n$ is said to be *absolutely convergent* if the series $\sum_n |a_n|$ converges. On the other hand, if $\sum_n a_n$ converges, but $\sum_n |a_n|$ diverges, then we say that $\sum_n a_n$ is *conditionally convergent*.

The example in the previous section shows that we cannot always expect to rearrange the terms of a conditionally convergent series without affecting the sum. We shall show that we *can* always do so for absolutely convergent series; first, however, we want to *relate* convergence of $\sum_n |a_n|$ to that of $\sum_n a_n$. We know already (from the above example) that the series $\sum_n a_n$ can converge *without* converging absolutely. We do, however, have the opposite implication:

● Proposition 5

Absolute convergence implies convergence.

PROOF
Suppose the series $\sum_{n\geq 1} |a_n|$ converges, with sum t. For $k \in \mathbb{N}$ set $s_k = \sum_{n=1}^{k} a_n$ and $t_k = \sum_{n=1}^{k} |a_n|$, so that $t = \lim_{k\to\infty} t_k$. Now $0 \leq a_n + |a_n| \leq 2|a_n|$, so that the series $\sum_{n\geq 1}(a_n + |a_n|)$ converges by comparison with $2\sum_{n\geq 1} |a_n|$. Write $u_k = \sum_{n=1}^{k}(a_n + |a_n|)$, so that the series $\sum_{n\geq 1}(a_n + |a_n|)$ has sum $u = \lim_{k\to\infty} u_k$. But $s_k = u_k - t_k$; therefore (s_k) converges to $s = u - t$, so that $\sum_{n=1}^{\infty} a_n = s$.

We can now see immediately that the Ratio and Root Tests are really tests for absolute convergence; we reformulate them as follows, using the above Proposition:

Ratio Test

Suppose we are given a real series $\sum_n a_n$.

(i) If there exists $r < 1$ and $n_0 \in \mathbb{N}$ such that $|\frac{a_{n+1}}{a_n}| \leq 1$ for all $n \geq n_0$ then $\sum_n a_n$ converges (absolutely).
(ii) If $|\frac{a_{n+1}}{a_n}| \geq 1$ for all n then $\sum_n a_n$ diverges.

Limit Form of the Test

Let $\sum_n a_n$ be a real series for which $\alpha = \lim_n |\frac{a_{n+1}}{a_n}|$ exists in $\mathbb{R}$. Then we have:

(i) if $\alpha < 1$ then $\sum_n a_n$ converges (absolutely);
(ii) if $\alpha > 1$ then $\sum_n a_n$ diverges;
(iii) if $\alpha = 1$ then the test is inconclusive.

Cauchy's nth *Root Test*

Let $\sum_n a_n$ be a real series. If there are $r < 1$ and $n_0 \in \mathbb{N}$ such that $\sqrt[n]{|a_n|} \le r$ for all $n > n_0$, then $\sum_n a_n$ converges (absolutely). If $\sqrt[n]{|a_n|} \ge 1$ for infinitely many n, then $\sum_n a_n$ diverges.

Upper Limit Form

Let $\sum_n a_n$ be a real series and let $\beta = \overline{\lim}_n \sqrt[n]{|a_n|}$. If $\beta < 1$, then $\sum_n a_n$ converges (absolutely) and if $\beta > 1$ then $\sum_n a_n$ diverges. The test is inconclusive when $\beta = 1$.

We return finally to the problem of *rearranging* series. For conditionally convergent series this can go badly wrong, as we have seen. In fact, it can be shown that a given conditionally convergent series can be rearranged to converge to *any* given limit (finite or infinite). We shall not prove this, but show instead that all is well as long as the series converges absolutely.

● *Theorem 7* ─────────────────────────────

If $\sum_n a_n$ converges absolutely, then so does any rearrangement of its terms, and has the same sum.

PROOF

We have not defined precisely what we mean by 'rearrangement' but the idea should be clear by now. First, consider the case where $a_n \ge 0$ for all n. Write $s = \sum_{n=1}^{\infty} a_n$ for the sum of this series, and let $\sum_k b_k$ be a rearrangement of the terms of $\sum_n a_n$, with partial sums (t_m), where $t_m = \sum_{k=1}^{m} b_k$. Then each t_m is the sum of *finitely many* of the a_n, so $t_m \le s$ for every $m \in \mathbb{N}$. Since the $b_k \ge 0$ this means that (t_m) is an increasing sequence, bounded above by s, and hence converges to $\sup_m t_m = t \le s$. On the other hand, we can exchange the roles of the two series: now if $\sum_k b_k$ is a series with positive terms and sum t, then it follows that the rearrangement $\sum_n a_n$ of its terms also converges and has sum $s \le t$. Hence $s = t$.

Now consider $a_n \in \mathbb{R}$ more generally: suppose that the real series $\sum_n a_n$ is absolutely convergent, that $\sum_n |a_n|$ has sum S, and that $\sum_k b_k$ is a rearrangement of $\sum_n a_n$. Then $\sum_k |b_k|$ is a rearrangement of the terms of $\sum_n |a_n|$. We can apply what we have just proved to the convergent series $\sum_n |a_n|$. So $\sum_k |b_k|$ converges to S also, i.e. $\sum_k b_k$ converges absolutely, as claimed.

Finally we need to show that the sums of $\sum_n a_n$ and $\sum_k b_k$ are equal in general. First, for any real number x, define $x^+ = \frac{1}{2}(|x| + x) = \max(x, 0)$ and $x^- = \frac{1}{2}(|x| - x) = \max(-x, 0)$. Note that x^+ and x^- are non-negative and that $x = x^+$ when $x > 0$, while $x = -x^-$ when $x < 0$. Thus $x = x^+ - x^-$ and $|x| = x^+ + x^-$ by definition.

Now the series $\sum_n a_n^+$, $\sum_n a_n^-$ converge by comparison with $\sum_n |a_n|$, and the series $\sum_k b_k^+$, $\sum_k b_k^-$ converge by comparison with $\sum_k |b_k|$, since all terms are non-negative. Moreover, $\sum_n a_n = \sum_n a_n^+ - \sum_n a_n^-$ and $\sum_k b_k = \sum_k b_k^+ - \sum_k b_k^-$, since all the series involved converge, so that we can apply Theorem 1 of Chapter 2 to the partial sums in each case. On the other hand, $\sum_k b_k^+$ is a rearrangement of $\sum_n a_n^+$, and similarly for $\sum_k b_k^-$ and $\sum_n a_n^-$. Hence, as these are series with positive terms, we know that $\sum_{n=1}^{\infty} a_n^+ = \sum_{k=1}^{\infty} b_k^+$ and $\sum_{n=1}^{\infty} a_n^- = \sum_{k=1}^{\infty} b_k^-$. Thus, finally, $\sum_{k=1}^{\infty} b_k = \sum_{n=1}^{\infty} a_n$.

EXERCISES ON 3.4

1. In each of the following cases decide whether the series converges absolutely or conditionally. Give reasons for your answers.

   (i) $\sum_{n\geq 1} \frac{\log n}{n^3}$ (ii) $\sum_{n\geq 1} (-1)^{n-1} (\frac{n+\sqrt{n}}{2n^3})$ (iii) $\sum_{n\geq 1} \frac{\sin \sqrt{n}}{n^{3/2}}$

   (iv) $\sum_{n\geq 1} \frac{(-1)^{n-1} n}{n^2+1}$ (v) $\sum_{n\geq 1} \frac{(-1)^{n-1} 2^n}{n^2}$ (vi) $\sum_{n\geq 1} n^2 e^{-n^2}$

2. Recall that if the series $\sum_n |a_n|$ converges, then so (by comparison) do the series $\sum_n a_n^+$ and $\sum_n a_n^-$ (where, for any $x \in \mathbb{R}$, $x^+ = \max(x, 0)$ and $x^- = \max(-x, 0)$). Now decide whether the converse is true, i.e. if both $\sum_n a_n^+$ and $\sum_n a_n^-$ converge, does it follow that $\sum_n a_n$ converges absolutely?

3. Suppose that $\sum_n a_n$ is convergent and $\sum_n b_n$ is divergent. Can $\sum_n (a_n + b_n)$ be convergent? Explain. (If you don't know how to start in general, try out $\sum_n \frac{1}{n}$ and $\sum_n \frac{1}{n^2}$ first.)

Summary

In this chapter we have consistently exploited the completeness property of the real number system to find methods for determining the limit of a convergent sequence, first for monotone bounded sequences, and then via an analysis of the behaviour of subsequences, concentrating on the consequences of the Bolzano–Weierstrass Theorem. Next, a number of tests for the convergence of series were derived, first for series with positive terms, and for series with alternating signs. We showed that the terms of the latter cannot always be rearranged without changing the sum. Finally, absolute convergence was introduced, and it was shown that any rearrangement of an absolutely convergent series converges to the same limit.

FURTHER EXERCISES

1. Investigate the convergence or divergence of the following series:

   (i) $1 + 2r + r^2 + 2r^3 + r^4 + 2r^5 + \dots$ when (a) $r = \frac{1}{2}$, (b) $r = \frac{3}{2}$.

   (ii) $\sum_{n\geq 1} n^{-\sqrt{n}} = 1 + \frac{1}{2^{\sqrt{2}}} + \frac{1}{3^{\sqrt{3}}} + \dots$

2. Find the sum of the following series:

   (i) $\sum_{n\geq 1} \frac{1}{n(n+3)}$

   (ii) $\frac{1}{1.3} + \frac{1}{3.5} + \frac{1}{5.7} + \dots$ (here . denotes multiplication).

3. Only one of the following statements is true. Use the comparison test to decide which it is, prove it and give an example to show that the other statement is false:

   (i) If $\sum_{n\geq 1} x_n^2$ converges then so does $\sum_{n\geq 1} |x_n|$.

   (ii) If $\sum_{n\geq 1} |x_n|$ converges then so does $\sum_{n\geq 1} x_n^2$.

4. Prove the following Quotient Test:

   Suppose $a_n \geq 0$ and $b_n \geq 0$ for all $n \geq 1$ and $\frac{a_n}{b_n} \to L$. Then:

   (i) if L is neither 0 nor ∞, then either both $\sum_n a_n$ and $\sum_n b_n$ converge or both diverge.

(ii) if $L = 0$ and $\sum_n b_n$ converges then so does $\sum_n a_n$.

(iii) if $L = \infty$ and $\sum_n b_n$ diverges then so does $\sum_n a_n$.

Using $b_n = \frac{1}{n^p}$ in the above, show that if $\lim_{n \to \infty} n^p a_n = L$ then we have: if $p > 1$ then $\sum_n a_n$ converges, while if $L \neq 0$ and $p \leq 1$ then $\sum_n a_n$ diverges.

Now investigate the convergence or divergence of the series $\sum_{n \geq 1} \frac{1}{n^{1+\frac{1}{n}}}$. How does your result relate to the convergence of $\sum_n \frac{1}{n^p}$ for $p > 1$?

Finally, use the Integral Test to show that $\sum_{n \geq 2} \frac{1}{n \log n}$ diverges but $\sum_{n \geq 2} \frac{1}{n(\log n)^2}$ converges. What does this remind you of?

5. *The Fibonacci sequence.* Let $a_1 = 1$, $a_2 = 2$, $a_{n+1} = a_n + a_{n-1}$ for $n \geq 2$. Show by induction that for all n

$$a_{n+1}a_{n+2} - a_n a_{n+3} = (-1)^n$$

Consider the ratios of successive terms: $b_n = \frac{a_{n+1}}{a_n}$ $(n \in \mathbb{N})$.

Show that $1 \leq b_n \leq 2$ for all n. Now show that each of the subsequences (b_{2n-1}) and (b_{2n}) is strictly monotone, and hence convergent. Show that their limits are the same. (*Hint:* consider Exercises on 3.1(5).)

Hence show that (b_n) converges to the golden section $\frac{1+\sqrt{5}}{2}$.

4 • Functions Defined by Power Series

4.1 Polynomials – and what Euler did with them!

We define a *polynomial* as a real function of the form $P_n(x) = a_0 + a_1 x + a_2 x^2 + a_3 x^3 + \ldots + a_n x^n$, where the (a_i) are fixed real numbers. If $a_n \neq 0$, we say that P_n has *degree n*. This gives rather a limited class of functions, although we shall see that polynomials can be used very effectively to approximate all sorts of more complicated functions. Polynomials have the advantage of being the simplest class of functions for the operations of the Calculus, since we can differentiate and integrate them term by term (as we shall prove in Chapters 8 and 10). Our immediate question, however, is what happens when we let $n \to \infty$: since for each fixed x, $P_n(x)$ is just a finite sum of real numbers, we can change our perspective slightly. We suppose that the sequence of coefficients $(a_i)_{i \geq 0}$ is given, and for each $x \in \mathbb{R}$ we regard the value $P_n(x)$ as the nth partial sum of a series. The partial sums for a particular x are then:

$$P_0(x) = a_0, \quad P_1(x) = a_0 + a_1 x, \quad P_2(x) = a_0 + a_1 x + a_2 x^2, \ldots$$

$$_n(x) = a_0 + a_1 x + a_2 x^2 + \ldots + a_n x^n, \ldots$$

Again we can ask for which $x \in \mathbb{R}$ the sum '$P_\infty(x)$' will make sense. To do this, we define a *power series* as a series of the form $\sum_{i \geq 0} a_i x^i$. This will make sense by our previous results whenever the sequence of partial sums $(P_n(x))_{n \geq 0}$ converges, and we can use the tests for convergence we developed in Chapter 3.

On the other hand, for all those x where the power series converges, the value of the sum $S(x) = \sum_{i=0}^{\infty} a_i x^i$ defines a value of a *real function S*. Power series thus provide a natural bridge leading us to the consideration of more general real functions. In fact, many of the common 'transcendental' functions, i.e. those whose definition 'transcends algebra' since they cannot be built as ratios or other algebraic combinations of polynomials, are nevertheless *limits* of polynomials in this natural way. We now examine the power series leading to such functions in more detail, concentrating on the well-known examples of the exponential and trigonometric functions, that is, we shall justify the *definitions*:

$$\exp(x) = \sum_{n=0}^{\infty} \frac{x^n}{n!} = 1 + \frac{x}{1!} + \frac{x^2}{2!} + \frac{x^3}{3!} + \ldots + \frac{x^n}{n!} + \ldots$$

$$\sin(x) = \sum_{n=0}^{\infty} (-1)^n \frac{x^{2n+1}}{(2n+1)!} = x - \frac{x^3}{3!} + \frac{x^5}{5!} - \ldots + (-1)^n \frac{x^{2n+1}}{(2n+1)!} + \ldots$$

$$\cos(x) = \sum_{n=0}^{\infty}(-1)^n \frac{x^{2n}}{(2n)!} = 1 - \frac{x^2}{2!} + \frac{x^4}{4!} - \dots + (-1)^n \frac{x^{2n}}{(2n)!} + \dots$$

as well as many others which follow from them.

Historical comment

In the eighteenth century, when the development of Calculus techniques led mathematicians to exploit the properties of power series (especially Taylor and Maclaurin series) which arose naturally in the Calculus, most mathematicians had few qualms about treating power series as 'true' polynomials, and using them as such. Aspects of these developments, including Lagrange's ill-fated attempt to use power series to *define* derivatives using power series, are recorded in [*NSS*], Section 5.2.

Example 1

Perhaps the most consistent advocate of treating infinite series as polynomials was Leonhard Euler (1707–83). Here is his remarkable calculation of the sum of the series $\sum_{n\geq 1} \frac{1}{n^2}$.

Fig. 4.1 Leonhard Euler (1707–83) (From the portrait by A. Lorgna)

Leonhard Euler was the most prolific mathematician of his time, producing over 800 books and papers. In later life he was blind and dictated his results to his sons. A series of texts he wrote in the 1750s remained the most influential textbooks on advanced calculus for nearly a century.

Euler begins his study of the series elsewhere: consider the quadratic equation $(x - a)(x - b) = 0$. Divide by ab to make the constant term 1. We obtain:

$$\frac{1}{ab}x^2 - \left(\frac{1}{a} + \frac{1}{b}\right)x + 1 = 0$$

We can read this as follows: 'if the constant term is 1, the linear term has coefficient equal to minus the sum of the reciprocals of the roots'. This simple algebraic fact can easily be extended to polynomials of degree n: if the polynomial equation

$$P_n(x) = a_0 + a_1 x + a_2 x^2 + \ldots + a_n x^n = 0$$

has roots $\alpha_1, \alpha_2, \ldots, \alpha_n$, then $-\frac{a_1}{a_0} = \frac{1}{\alpha_1} + \frac{1}{\alpha_2} + \ldots + \frac{1}{\alpha_n}$.

Euler applies this directly to the 'polynomial' $\sin(x) = x - \frac{x^3}{3!} + \frac{x^5}{5!} - \frac{x^7}{7!} + \ldots$, which has roots $x = n\pi$ ($n \in \mathbb{Z}$). First divide by x to remove the root $x = 0$, so that

$$\frac{\sin x}{x} = 1 - \frac{x^2}{3!} + \frac{x^4}{5!} - \frac{x^6}{7!} + \ldots = 0$$

has roots $\pm n\pi$ ($n \in \mathbb{N}$). Finally, let $u = x^2$, so that

$$1 - \frac{u}{3!} + \frac{u^2}{5!} - \frac{u^3}{7!} + \ldots = 0$$

has roots $n^2\pi^2$ ($n \in \mathbb{N}$). Euler's 'theorem' says that the sum of the reciprocals of these roots, i.e.

$$\frac{1}{\pi^2} + \frac{1}{4\pi^2} + \frac{1}{9\pi^2} + \ldots + \frac{1}{n^2\pi^2} + \ldots$$

should equal the coefficient of the linear term, i.e. $-(-\frac{1}{3!}) = \frac{1}{6}$. In other words, $\sum_{n=1}^{\infty} \frac{1}{n^2} = \frac{\pi^2}{6}$.

The result of this calculation is correct, even though the argument rests on the unproven assertion that the algebraic result about the roots of a polynomial extends to 'polynomials' of infinite degree, such as $\sin(x)$. The modern substitute for this calculation is more elaborate, using results about Fourier series. The example serves to illustrate Euler's genius in arriving at correct results, even if the methods used were not fully justified at the time.

TUTORIAL PROBLEM 4.1

Another example of Euler's skill in handling infinite series is provided by his summing of the series

$$1 - \frac{1}{2} + \frac{1}{3} + \frac{1}{4} - \frac{2}{5} + \frac{1}{6} + \frac{1}{7} + \frac{1}{8} - \frac{3}{9} + \ldots$$

Euler calculated the sum as if he were handling a series with finitely many terms: discuss the outline of his argument given below, and decide for yourself whether you believe the result. Deriving it directly with the tools you now have is distinctly tricky!

Euler first had to decide what the general term looks like: the denominators increase steadily by 1, and all numerators are 1, except those for numbers of the form $\frac{j(j+1)}{2} - 1$ (i.e. the numbers which immediately precede triangular numbers!), which are given the numerator $-(j-1)$. Euler treated the series as having $N(\frac{N+3}{2}) = \frac{K}{2}(K+1) - 1$ (where $K = N + 1$) terms, where N is infinite. Euler stopped the series at a term where the cycle is complete, i.e. where the final term has negative sign. Thus he could treat the series as the difference of two series (both divergent, of course!) namely:

$$1 + \frac{1}{2} + \frac{1}{3} + \ldots + \frac{1}{N(\frac{N+3}{2})}$$

and

$$\frac{2}{2} + \frac{3}{5} + \frac{4}{9} + \frac{5}{14} + \ldots + \frac{N+1}{N(\frac{N+3}{2})}$$

Now he applied his formula for the sum of the (divergent) harmonic series, using 'Euler's constant' $\gamma = \lim_{n\to\infty}[(\sum_{k=1}^{n}\frac{1}{k}) - \log n]$, which we derived in Section 3.3. Applying this with infinite N he obtained the sum

$$\gamma + \log\left(N\frac{N+3}{2}\right) \approx \gamma + \log N + \log(N+3) - \log 2$$

for the first series, where $\approx$ indicates that the two sides are infinitely close to each other. The second series was then decomposed into partial fractions:

$$\frac{2}{3}\left(1 + \frac{1}{2} + \frac{1}{3} + \ldots + \frac{1}{N}\right) + \frac{4}{3}\left(\frac{1}{4} + \frac{1}{5} + \ldots + \frac{1}{N+3}\right)$$

which gave the sums $\frac{2}{3}(\gamma + \log N)$ and $\frac{4}{3}(\gamma + \log(N+3) - 1 - \frac{1}{2} - \frac{1}{3})$ respectively. Subtracting, he claimed finally that our original series has sum $\frac{22}{9} - \gamma - \log 2$.

4.2 Multiplying power series: Cauchy products

We have seen that term-by-term addition, and more generally, forming linear combinations of terms, presents no problems when we discuss convergence of series in general. On the other hand, multiplication of series, just as for polynomials, leads one quite naturally to rearrangements of the order in which the terms are considered, and this becomes problematical when the terms are no longer necessarily positive. We now tackle this question for the particular case of power series. The gist of our discussion of rearrangements of series in Chapter 3 was that care must be taken in general, but that all is well when the series in question converge *absolutely*. This insight will prove useful below.

When we multiply two finite sums of terms $(a_0 + a_1 + a_2)$ and $(b_0 + b_1 + b_2)$ we can, for instance, write the products in the following order:

$$(a_0 + a_1 + a_2)(b_0 + b_1 + b_2) =$$
$$a_0b_0 + (a_0b_1 + a_1b_0) + (a_0b_2 + a_1b_1 + a_2b_0) + (a_1b_2 + a_2b_1) + a_2b_2$$

The nine terms of this product have been grouped as they would be when we multiply the *polynomials* $(a_0 + a_1x + a_2x^2)$ and $(b_0 + b_1x + b_2x^2)$: then each group would correspond to the coefficient of a power of x, arranged in increasing order of these powers. Hence to extend multiplication of polynomials to that of power series $(\sum_m a_m x^m)(\sum_n b_n x^n)$, we have a 'natural' product suggested by polynomial multiplication:

● Definition I

The *Cauchy product* of the power series $\sum_{m \geq 0} a_m x^m$ and $\sum_{n \geq 0} b_n x^n$ is the series $\sum_{k \geq 0} c_k x^k$, where for each $k \geq 0$,

$$c_k = a_0 b_k + a_1 b_{k-1} + \ldots + a_k b_0 \quad \text{(Note: } k+1 \text{ terms.)}$$

● Theorem I ———————————————————

Suppose that $\sum_{m \geq 0} a_m x^m$ and $\sum_{n \geq 0} b_n x^n$ are both absolutely convergent, with sums $f(x)$ and $g(x)$ respectively. Then their Cauchy product $\sum_k c_k x^k$ is absolutely convergent and its sum equals $f(x).g(x)$.

Actually, we can do rather more: given *any* two series $\sum_{m \geq 0} a_m$ and $\sum_{n \geq 0} b_n$ we can define their Cauchy product as $\sum_{k \geq 0} c_k$, where $c_k = \sum_{i=0}^{k} a_i b_{k-i}$, just as above. (Power series are just a special case.) The Cauchy product of absolutely convergent series is also absolutely convergent as we now show:

● Proposition I

Suppose that $\sum_{m=0}^{\infty} a_m = s$, $\sum_{n=0}^{\infty} b_n = t$ and assume that both series converge absolutely. Then their Cauchy product is absolutely convergent and has sum st.

PROOF
Suppose we have proved this for series with positive terms. Then we can use the convergence of $\sum_m |a_m|$ and $\sum_n |b_n|$ to deduce that $\sum_k c_k$ converges absolutely. As in Theorem 7 in Chapter 3 we can then use the fact that $a_m = a_m^+ - a_m^-$ and $b_n = b_n^+ - b_n^-$ and the validity of our claim for series with positive terms to show that $\sum_{k=0}^{\infty} c_k = st$. Hence we can assume without loss that our series have positive terms.
Now we only need to show that $\sum_k c_k$ converges to the sum st. For this, write the possible products in a 'matrix' array as follows:

$$
\begin{array}{ccccc}
a_0 b_0 & a_0 b_1 & a_0 b_2 & a_0 b_3 & \cdots \\
a_1 b_0 & a_1 b_1 & a_1 b_2 & a_1 b_3 & \cdots \\
a_2 b_0 & a_2 b_1 & a_2 b_2 & a_2 b_3 & \cdots \\
a_3 b_0 & a_3 b_1 & a_3 b_2 & a_3 b_3 & \cdots \\
\cdots & \cdots & \cdots & \cdots & \cdots
\end{array}
$$

Now note that, if $s_k = \sum_{m=0}^{k} a_m$ and $t_k = \sum_{n=0}^{k} b_n$ denote the kth partial sums of the original series, and if we write u_k for the sum of all the products $a_i b_j$ with $i, j \leq k$, then we obtain:

$$u_0 = a_0 b_0 = s_0 t_0$$

$$u_1 = a_0 b_0 + a_1 b_0 + a_1 b_1 + a_0 b_1 = (a_0 + a_1)(b_0 + b_1) = s_1 t_1$$

and, in general,

$$u_k = (a_0 + a_1 + \ldots + a_k)(b_0 + b_1 + \ldots + b_k) = s_k t_k$$

By the algebra of limits, $\lim_{k \to \infty} u_k = st$. On the other hand, the partial sums $v_j = \sum_{k=0}^{j} c_j$ of the Cauchy product $\sum_k c_k$ are obtained by only summing the diagonals of the array, that is,

$$a_0 b_0 + (a_0 b_1 + a_1 b_0) + (a_0 b_2 + a_1 b_1 + a_2 b_0) + \ldots$$

so that $v_j \le u_j$ for each j. Since the sum of the first $2j$ diagonals always includes all the products in the first $j \times j$ square in the array, it is clear that $u_j \le v_{2j}$ for each j. Thus the terms of the sequence (v_j) are sandwiched between terms of (u_j) and it follows that they have the same limit. This proves the Proposition.

TUTORIAL PROBLEM 4.2

(i) In order to show only that $\sum_k c_k$ converges to st, we need only *one* of $\sum_m a_m$ and $\sum_n b_n$ to converge absolutely. Using the notation of the above proof, assume that $\sum_m a_m$ converges absolutely, and write $w_n = t_n - t$. Show first that we can write the partial sum v_n in the form $v_n = s_n t + x_n$, where

$$x_n = a_0 w_n + a_1 w_{n-1} + \ldots + a_n w_0$$

Next, writing $x = \sum_{n=0}^{\infty} |a_n|$ (since we assume that $\sum_n a_n$ converges absolutely), use the fact that $w_n \to 0$ to prove that $x_n \to 0$ also. Hence conclude that $\sum_{k=0}^{\infty} c_k = st$.

(ii) It is possible to say still more: without any assumptions about absolute convergence, one can show that whenever the series $\sum_k c_k$ converges, its sum *must* be st. However, it is possible for $\sum_m a_m$ and $\sum_n b_n$ to converge *without* being able to conclude that $\sum_k c_k$ converges! An example of such series is given in Exercise on 4.2.

Example 2

We return to power series, and apply Theorem 1. In particular, for $|x| < 1$, the power series $\sum_{n \ge 0} x^n$ is absolutely convergent, and its Cauchy product with itself is $\sum_{k \ge 0} (k+1) x^k$. To see this note that $a_n = b_n = 1$ for all $n \ge 1$, so that $c_k = 1.1 + 1.1 + \ldots + 1.1 = k + 1$, as there are $k + 1$ terms in the above definition of c_k. On the other hand $\sum_{n=0}^{\infty} x^n = \frac{1}{1-x}$, so we have shown that $\sum_{n \ge 1} n x^{n-1} = \sum_{k \ge 0} (k+1) x^k = \frac{1}{(1-x)^2}$ whenever $|x| < 1$.

(This suggests strongly that 'term-by-term' differentiation has a future for power series ...)

Example 3

Similarly, as $\sum_{m\geq0}\frac{x^m}{m!}$ converges absolutely for all $x \in \mathbb{R}$, we can explore the Cauchy product $(\sum_{m=0}^{\infty}\frac{x^m}{m!})(\sum_{n=0}^{\infty}\frac{y^n}{n!})$ for general $x, y \in \mathbb{R}$. The Cauchy product is $\sum_{k=0}^{\infty}c_k$, where $c_k = \sum_{i=0}^{k}\frac{x^i}{i!}\frac{y^{k-i}}{(k-i)!} = \frac{1}{k!}\sum_{i=0}^{k}\binom{k}{i}x^iy^{k-i} = \frac{(x+y)^k}{k!}$ for each k. We have verified that, if we *define* the exponential function by: $\exp x \equiv \sum_{n=0}^{\infty}\frac{x^n}{n!}$ then it satisfies the identity $\exp x.\exp y = \exp(x+y)$ for all $x, y \in \mathbb{R}$.

EXERCISE ON 4.2

1. Show that the Cauchy product of the series $\sum_{n\geq0}(-1)^n\frac{1}{\sqrt{n+1}}$ with itself is given by the series $\sum_{n\geq0}c_n$, where $c_n = (-1)^n\sum_{k=0}^{n}\frac{1}{\sqrt{(n-k+1)(k+1)}}$. Prove that $\sqrt{(n-k+1)(k+1)} \leq \frac{1}{2}(n+2)$, and hence that $|c_n| \geq 1$ for all n. Why does this verify the claim made at the end of Tutorial Problem 4.2?

4.3 The radius of convergence of a power series

A natural question (already referred to for the case of the geometric series) is whether we can still differentiate and integrate such series *term-by-term*, as we do for polynomials. To answer this, we must first decide under what conditions (and for which x) the power series converges. Fortunately the answer is very precise: we can identify an *interval of convergence with radius R*, that is, a set of real numbers $(-R, R) = \{x \in \mathbb{R} : -R < x < R\}$ on which a given power series will converge absolutely.

● Theorem 2—Radius of convergence ────────────

Let $\sum_{n\geq0}a_nx^n$ be a real power series. The set of $x \in \mathbb{R}$ for which the series converges is an interval with midpoint 0.

More precisely, there are three possibilities: either the series converges only for $x = 0$, or it converges for all real x, or there exists $R \in \mathbb{R}$ such that $\sum_{n\geq0}a_nx^n$ converges absolutely whenever $|x| < R$ and diverges whenever $|x| > R$.

(We call R the *radius of convergence* of the series $\sum_{n\geq0}a_nx^n$. Note carefully that nothing is said about the behaviour of $\sum_{n\geq0}a_nx^n$ when $|x| = R$.)

PROOF
A very short proof uses the Upper Limit Form of the Root Test, applied to the sequence $x_n = |a_nx^n| :$ we have

$$\beta = \limsup_{n\to\infty}\sqrt[n]{x_n} = |x|\limsup_{n\to\infty}\sqrt[n]{|a_n|}$$

Let $\alpha = \limsup_n\sqrt[n]{|a_n|}$ and $R = \frac{1}{\alpha}$ (when $\alpha = 0$ we set $R = \infty$ and when $\alpha = \infty$ set $R = 0$). Then we have $\beta = \frac{|x|}{R}$, and $\sum_n x_n$ converges if $\beta < 1$ (i.e. $|x| < R$) and diverges if $\beta > 1$ (i.e. $|x| > R$). Thus $\sum_n a_nx^n$ converges absolutely when $|x| < R$ and diverges when $|x| > R$. As with the root test, the examples $\sum_n\frac{1}{n}$ and $\sum_n\frac{1}{n^2}$

show that we can deduce nothing when $|x| = R$: the power series $\sum_n \frac{x^n}{n}$ and $\sum_n \frac{x^n}{n^2}$ both have radius of convergence 1, but the first diverges as $x = 1$ while the second converges there.

The proof provides an explicit formula for the radius of convergence, known as *Hadamard's formula*:

$$R = \frac{1}{\limsup_n \sqrt[n]{|a_n|}}$$

TUTORIAL PROBLEM 4.3

Read [*NSS*] Section 8.5 for an alternative proof of Theorem 2, without using upper or lower limits. (Our proof, though already much shorter, will find further use when we look at derived series below.)

Example 4

The above formula is not a very practical method for finding R. Fortunately, there is often a very simple way to calculate the radius of convergence, using the Ratio Test instead.

(i) For example, if $\sum_{n \geq 0} \frac{(n!)^2}{(2n)!} x^n$ then the ratios of successive terms are

$$\left| \frac{((n+1)!)^2}{(2(n+1))!} x^{n+1} \right| \times \left| \frac{(2n)!}{(n!)^2 x^n} \right| = |x| \left(\frac{(n+1)(n+1)}{(2n+1)(2n+2)} \right)$$

and this converges to $\frac{1}{4}|x|$. The limit of the ratios is therefore less than 1 if and only if $|x| < 4$. Thus $R = 4$.

The next two series illustrate extreme cases:

(ii) $\sum_n n! x^n$ has radius of convergence 0, since the ratios of successive terms are $\frac{(n+1)!}{n!}|x| = (n+1)|x|$, which becomes greater than 1 for all $x \neq 0$ as $n \to \infty$.

(iii) On the other hand, $\sum_n \frac{x^n}{n!}$ converges for all real x, since the ratios are $\frac{|x|}{n+1}$, which converges to 0 for every $x \in \mathbb{R}$. (Recall that we write $R = \infty$ in this case.)

EXERCISES ON 4.3

1. Find the radius of convergence of the following power series:

   (i) $\sum_{n \geq 0} \frac{x^n}{3^n}$ (ii) $\sum_{n \geq 0} 4^n x^{2n}$ (iii) $\sum_{n \geq 0} (2n^3 - n^2) x^n$

   (iv) $\sum_{n \geq 1} (\frac{1}{2} + \frac{1}{n})^n x^{2n}$ (v) $\sum_{n \geq 0} (\sqrt{n+1} - \sqrt{n}) x^n$ (vi) $\sum_{n \geq 0} \frac{n!(2n)!}{(3n)!} x^n$

2. Determine the radius of convergence of the power series $\sum_{n \geq 1} n^r x^n$, where $r \in \mathbb{Q}$ is fixed. Hence show that for each $r \in \mathbb{Q}$ the sequence $(n^r x^n)_{n \geq 1}$ is null whenever $|x| < 1$.

4.4 The elementary transcendental functions

We are ready to describe in more detail how power series can be used to define the exponential, hyperbolic and trigonometric functions.

◈ *Example 5*

The following definitions have been given already in Section 4.1. We can check very easily that in each case the series converges absolutely for all real x (that is, $R = \infty$).

$$\exp(x) = \sum_{n=0}^{\infty} \frac{x^n}{n!} = 1 + \frac{x}{1!} + \frac{x^2}{2!} + \frac{x^3}{3!} + \ldots + \frac{x^n}{n!} + \ldots$$

$$\sin(x) = \sum_{n=0}^{\infty} (-1)^n \frac{x^{2n+1}}{(2n+1)!} = x - \frac{x^3}{3!} + \frac{x^5}{5!} - \ldots + (-1)^n \frac{x^{2n+1}}{(2n+1)!} + \ldots$$

$$\cos(x) = \sum_{n=0}^{\infty} (-1)^n \frac{x^{2n}}{(2n)!} = 1 - \frac{x^2}{2!} + \frac{x^4}{4!} - \ldots + (-1)^n \frac{x^{2n}}{(2n)!} + \ldots$$

It is clear that $\cos(-x) = \cos(x)$ for any x, since the terms in the defining series involve only *even* powers of x, whereas $\sin(-x) = -\sin(x)$ since here the defining series contains only *odd* powers of x. We say that a real function f is *even* if $f(-x) = f(x)$, and *odd* if $f(-x) = -f(x)$ for all real x. Thus cos is even and sin is odd. By contrast, the exponential function is neither even nor odd, since $\exp(x) = \sum_{n=0}^{\infty} \frac{x^n}{n!}$ involves all powers of x.

However, given any real function f we can 'split' f into the sum of an even and an odd function, i.e. $f = g + h$, with g even and h odd, simply by taking $g(x) = \frac{1}{2}(f(x) + f(-x))$ and $h(x) = \frac{1}{2}(f(x) - f(-x))$.

In particular, apply this with $f(x) = \exp(x)$: thus we define $g = \cosh$, $h = \sinh$, by: $\cosh(x) = \frac{1}{2}(\exp(x) + \exp(-x))$, $\sinh(x) = \frac{1}{2}(\exp(x) - \exp(-x))$.

In terms of the series defining exp this becomes:

$$\cosh(x) = \sum_{n=0}^{\infty} \frac{x^{2n}}{(2n)!} \qquad \sinh(x) = \sum_{n=0}^{\infty} \frac{x^{2n+1}}{(2n+1)!}$$

Here we have merrily added and subtracted power series as if they were polynomials. But this follows from a general fact:

The *arithmetic* of power series can be deduced from that for general absolutely convergent series: to be precise, let $f(x) = \sum_{n=0}^{\infty} a_n x^n$ ($|x| < R_1$) and $g(x) = \sum_{n=0}^{\infty} b_n x^n$ ($|x| < R_2$) be functions defined by power series whose radii of convergence are R_1 and R_2 respectively. Then f and g are well-defined real functions at least on the smaller of the two intervals of convergence (we can suppose $0 < R_1 < R_2$ for definiteness), and we have immediately that, for $\alpha, \beta \in \mathbb{R}$:

$$\alpha f(x) + \beta g(x) = \alpha \sum_{n=0}^{\infty} a_n x^n + \beta \sum_{n=0}^{\infty} b_n x^n = \sum_{n=0}^{\infty} (\alpha a_n + \beta b_n) x^n$$

and $f(x)g(x) = (\sum_{n=0}^{\infty} a_n x^n)(\sum_{n=0}^{\infty} b_n x^n) = \sum_{n=0}^{\infty} c_n x^n$, where the right-hand side is the Cauchy product, i.e. $c_n = \sum_{k=0}^{n} a_k b_{n-k}$ for each n.

TUTORIAL PROBLEM 4.4

Recall the definitions of $\cos(x) = \sum_{n=0}^{\infty}(-1)^n \frac{x^{2n}}{(2n)!}$ and $\sin(x) = \sum_{n=0}^{\infty}(-1)^n \frac{x^{2n+1}}{(2n+1)!}$, which are valid for every real x, since the series converge absolutely.

Now use Cauchy products to prove that

$$\cos^2(x) = \sum_{n=0}^{\infty} \alpha_n x^{2n} \qquad \sin^2(x) = \sum_{n=0}^{\infty} \beta_n x^{2n+2}$$

where for each $n \in \mathbb{N}$

$$\alpha_n = \sum_{k=0}^{n} \frac{(-1)^k}{(2k)!} \frac{(-1)^{n-k}}{(2n-2k)!} \qquad \beta_n = \sum_{l=0}^{n} \frac{(-1)^l}{(2l+1)!} \frac{(-1)^{n-l}}{(2n-2l+1)!}$$

Next, show that $\cos^2(x) + \sin^2(x) = 1 + \sum_{n=0}^{\infty}(\alpha_n + \beta_{n-1})x^{2n}$. Finally, use the binomial expansion of $(1+x)^{2n}$ with $x = -1$ to show that the last sum is in fact 0, thus proving the familiar relation $\cos^2(x) + \sin^2(x) = 1$. (We shall give a *much* less arduous proof of the same result in Chapter 9!)

Note that in all the above examples the function *values f(x)* and *g(x)* are given as sums of convergent series of real numbers, in fact, as limits of partial sums: $f(x) = \lim_{k \to \infty} s_k(x)$, where $s_k(x) = \sum_{n=0}^{k} a_n x^n$ and similarly for $g(x)$. This holds for every $x \in (-R_1, R_1)$, and $x \mapsto s_k(x)$ is a polynomial defined on this interval. In order to show that the above definitions of the functions exp, sin, cos, sinh, cosh have really given us a rigorous way of treating the *analysis* of these functions we shall need to show that the elementary analytic properties of polynomials can be 'transplanted' to infinite sums as well. To do this, we shall need to decide what these 'analytic properties' of *real functions* should be. That is the content of the rest of this book: for now we can give a foretaste of what is needed, using Hadamard's formula once again.

Clearly if we want to develop foundations for the Calculus, these properties will need to include the ability to differentiate and integrate – for power series we would certainly expect to do this term by term. To see that such hopes may well be realized we now show that the series we would *expect* to consider at least have the same radius of convergence.

● Proposition 2

The three power series

$$\sum_{n \geq 0} a_n x^n \qquad \sum_{n \geq 1} n a_n x^{n-1} \qquad \sum_{n \geq 0} \frac{a_n}{n+1} x^{n+1}$$

all have the same radius of convergence.

PROOF

Suppose that the original series $\sum_n a_n x^n$ has radius of convergence R. Then $R = \frac{1}{\alpha}$, where $\alpha = \limsup_n \sqrt[n]{|a_n|}$. But since $|na_n|^{\frac{1}{n}} = n^{\frac{1}{n}} |a_n|^{\frac{1}{n}}$ and

$$| \frac{a_n}{n+1} |^{\frac{1}{n}} = \frac{1}{(n+1)^{\frac{1}{n}}} |a_n|^{\frac{1}{n}}$$

it follows from the fact that $n^{\frac{1}{n}} \to 1$ as $n \to \infty$, that

$$\alpha = \limsup_n \sqrt[n]{|na_n|} = \limsup_n \sqrt[n]{|\frac{a_n}{n+1}|}$$

Hence the result follows.

Example 6

Consider the exponential function again: then $a_n = \frac{1}{n!}$, so that $na_n = \frac{1}{(n-1)!}$ for $n \geq 1$. By Proposition 2 the series $\sum_{n \geq 1} na_n x^{n-1}$ and $\sum_{n \geq 0} \frac{a_n}{n+1} x^{n+1}$ both converge for all x. But in this case the first of these series is *identical* with $\sum_{n \geq 0} a_n x^n$, since $na_n x^{n-1} = \frac{x^{n-1}}{(n-1)!}$, so that

$$\sum_{n \geq 1} na_n x^{n-1} = \sum_{n \geq 1} \frac{x^{n-1}}{(n-1)!} = \sum_{m \geq 0} \frac{x^m}{m!}$$

Similarly $\frac{a_n}{n+1} x^{n+1} = \frac{x^{n+1}}{(n+1)!}$ so that $\sum_{n \geq 0} \frac{a_n}{n+1} x^{n+1} = \sum_{m \geq 1} \frac{x^m}{m!}$, which differs from the original series $\sum_{m \geq 0} \frac{x^m}{m!}$ by the additive constant a_0. Of course these two series are just what we would obtain from the series for exp by term-by-term differentiation and integration, respectively. What we have shown is that this process would yield the 'well-known fact' that exp 'is its own derivative and indefinite integral'.

Thus our definitions of the basic transcendental functions have the right 'shape' to fit with the Calculus results we wish to verify. What is now needed are theorems which justify term-by-term differentiation and integration. At the same time, the fact that we resort to power series as 'polynomials of infinite degree' in order to define these new functions suggests that polynomials should play a central role in the analysis of the classes of functions we shall now investigate: the question whether 'every decent function' can be *approximated* by polynomials has a long history, and is the basis of many practical applications of the Calculus.

EXERCISE ON 4.4

1. Use the series definition of exp to deduce the following simple asymptotic properties of this function:
   (i) $\exp x \to +\infty$ as $x \to +\infty$
   (ii) $\exp x \to 0$ as $x \to -\infty$
   (iii) $\lim_{x \to +\infty} x^n \exp(-x) = 0$ for all $n \in \mathbb{N}$.
   (The final result shows that 'exponentials dominate powers'.)

Summary

Beginning with polynomials, we have described the natural arithmetic properties of power series, which allow us to use such series to define functions within specified domains. The fact that power series converge *absolutely* inside their radius of convergence ensures that the terms of the series can be rearranged at will without altering their convergence or sum. This allows one to define the Cauchy product of two power series as a natural extension of the multiplication of polynomials, and yields some of the familiar properties of exponential and trigonometric functions – though these must now be treated as functions *defined* by their power series expansions.

FURTHER EXERCISES

1. Find the radius of convergence of the following power series:

    (i) $\sum_{n\geq 0} \frac{1+2^n}{1+3^n} x^n$ (ii) $\sum_{n\geq 1}(-1)^n \frac{nx^n}{1+n^2}$

2. Recall that the binomial coefficient of a real number a which is neither 0 nor a positive integer, can be defined as $\binom{a}{n} = \frac{a(a-1)(a-2)...(a-n+1)}{n!}$ for all $n \in \mathbb{N}$. Show that the power series $\sum_{n\geq 0} \binom{a}{n} x^n$ has radius of convergence 1.

3. Suppose the power series $\sum_{n\geq 0} a_n x^n$ has radius of convergence 1 and hence defines the function $f(x) = \sum_{n=0}^{\infty} a_n x^n$ for $|x| < 1$. Use a Cauchy product to show that $f(x)(1-x)^{-1} = \sum_{n\geq 0} b_n x^n$, where $b_n = a_0 + a_1 + \ldots + a_n$ for all n.

4. Use the previous two Exercises to prove the following identity, which is known as the binomial theorem for a negative integer exponent:

    for $m \in \mathbb{N}$ and $|x| < 1$

    $$(1-x)^{-m} = 1 + mx + \binom{m}{2}x^2 + \ldots + \binom{m}{n}x^n + \ldots$$

    where the series on the right converges absolutely. (*Hint*: use induction on m.)

5 • Functions and Limits

Although we have already considered several types of 'real function', such as polynomials and the functions defined via power series, we now embark on a more detailed study of the analytical properties of general functions whose domain and range are both contained in $\mathbb{R}$. Thus we first need to define precisely what we mean by such a *real function*, so that we can then describe the intuitive notion of the 'limit of $f(x)$ as x tends to a' and other concepts, such as *continuity, differentiability and integrability*, which occupied much of the time and attention of many of the most famous names of eighteenth and nineteenth century mathematics. Unlike these, we shall base our description of function on the concept of *sets*, introduced only late in the nineteenth century by Georg Cantor (1845–1918).

Fig. 5.1 Georg Cantor (1845–1918)

The German mathematician Georg Cantor was prominent in the late nineteenth century search for foundations for Analysis; his investigations into Fourier series led him to a closer examination of the 'sizes' of infinite collections of points, especially those which are unevenly distributed on the real line. His theory of ordinal and cardinal numbers opened up an entirely new world of abstract sets, though much of his interest remained in exploring the nature of the real number line. His work was controversial, and led to virulent attacks on him by some of his contemporaries, notably his former colleague in Berlin, Leopold Kronecker. Cantor suffered several nervous breakdowns and spent his last years in an asylum.

5.1 Historical interlude: curves, graphs and functions

The description of a curve by a *formula*, such as those of the parabola centred at the origin ($y = ax^2$), the circle centred at (x_0, y_0) and with radius r (i.e. $(x - x_0)^2 + (y - y_0)^2 = r^2$), or the cissoid of Diocles (i.e. $y^2 = (x^2 + y^2)x$) goes back to the invention of a system of *coordinate geometry* by the French philosopher and mathematician René Descartes in 1637. This development ushered in the 'golden age' of seventeenth century mathematics which saw the flowering of the Calculus, developed by many contributors and perfected in the period 1660–90, first by Isaac Newton (1642–1727) and then independently by Gottfried Wilhelm Leibniz (1646–1716).

Fig. 5.2   René Descartes (1596–1650) (From a painting by Frans Hals)

While Newton, Leibniz and their contemporaries *began* with the geometrical figures described by the Ancient Greeks, they quickly found that the new *algebraic* formulae by which Descartes' methods allowed them to describe and analyse such figures, could be extended with ease to 'figures' whose geometric description was much less obvious. In time, the geometric shape of the 'curves' became secondary to the formulae describing the *relationship* between the coordinates (x, y) of the 'moving point' which traces out the figure in question. The Calculus appeared to

Fig. 5.3 Isaac Newton (1642–1727) (From a portrait by Godfrey Kneller)

Fig. 5.4 Gottfried Wilhelm Leibniz (1646–1716) (From a picture in the Uffizi Gallery, Florence)

apply not just to geometrical figures, but to a much wider class of *functions* which could be defined by the 'rules' or formulae which relate these coordinates to each other.

By the end of the eighteenth century the rapid development of the Calculus and its many applications had focused much attention on the classes of functions which one could integrate and on the meaning of notions of *limit*, *derivative* and *continuity.* (*See* [NSS], Chapter 5 for a brief discussion of where this search for foundations was to lead.) Since many applications demanded the consideration of an ever wider class of 'formulae' the concept of *function* was gradually extended, so that ultimately it lost touch with its geometric origins.

By the middle of the nineteenth century, functions were accepted which were no longer defined by a single formula, but which required different formulae in different parts of their domain: among these was the humble *modulus* function $f(x) = |x| = \left\{ \begin{smallmatrix} x & \text{if } x \geq 0 \\ -x & \text{if } x < 0 \end{smallmatrix} \right.$, as well as functions whose graphs could no longer be drawn at all accurately, such as the *indicator function* $1_{\mathbb{Q}}$ of the set of the set $\mathbb{Q}$ of all rational numbers, i.e.

$$1_{\mathbb{Q}}(x) = \begin{cases} 1 & \text{if } x \in \mathbb{Q} \\ 0 & \text{otherwise} \end{cases}$$

and the function g defined by the formulae

$$g(x) = \begin{cases} 0 & \text{if } x = 0 \\ \frac{1}{n} & \text{if } \frac{1}{n} \leq |x| < \frac{1}{n-1} \ (n \in \mathbb{N}) \end{cases}$$

TUTORIAL PROBLEM 5.1

Actually, $f(x) = |x|$ *does* have a description by a single formula for all $x \in \mathbb{R}$. Discuss this description, and also explain the 'sketch' of g given below.

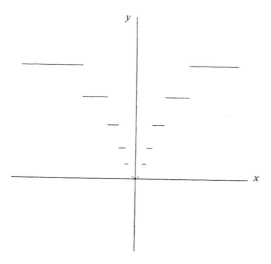

Fig. 5.5 $g(x) = \left\{ \begin{smallmatrix} 0 \\ 1/n \end{smallmatrix} \right.$ $\begin{smallmatrix} \text{if } x=0 \\ \text{if } \frac{1}{n} \leq |x| < \frac{1}{n-1} \end{smallmatrix}$

An even stranger example is the bizarre *Dirichlet* function

$$h(x) = \begin{cases} \frac{1}{n} & \text{if } x = \frac{m}{n} \in \mathbb{Q} \\ 0 & \text{if } x \notin \mathbb{Q} \end{cases}$$

whose properties we shall examine in Chapter 6. (You may wish to attempt a rough sketch now, though!)

Moreover, an ever larger class of functions which were defined via power series and other series of functions began to become important in the solution of differential equations – these could not simply be given a 'closed form expression', but had to be handled separately.

It also became clear that the *existence* of a 'formula' does not, by itself, define a meaningful function: consider the formula $y = \log(\log(\sin x))$ and ask yourself for which real numbers x this makes sense! As early as the 1750s Leonhard Euler (1707–83) had revised his earlier view that a function should always be defined via an *algebraic* formula linking the x- and y-coordinates: he now argued that *any* 'rule' *or set of rules* fixing y for each particular x in question should be treated as giving a y as a function of x. The task of making sense of the 'limiting behaviour' of such functions, and of interpreting the operations of the Calculus in this general framework, led nineteenth century mathematicians such as Cauchy, Dirichlet, Riemann and Weierstrass to subtle problems of the *classification* of functions and also describing more precisely the underlying system of *real numbers* on which such functions are defined.

5.2 The modern concept of function: ordered pairs, domain and range

Since the advent of set theory we can be much bolder still: if we accept that a function simply describes the relationship between two 'variable quantities' then we need not restrict ourselves to real numbers, but can simply consider correspondences between the elements of two *sets* A and B. The function $f : A \mapsto B$ tells us which pairs (a, b) correspond to each other: we write $b = f(a)$ simply to denote this correspondence. Thus the only purpose of any 'rule' is to describe *which sets of pairs* (a, b) we wish to consider – the 'rule' itself is therefore unnecessary, and we consider the set of pairs directly. Thus a function can be defined very precisely as a certain kind of *set*. It is a particular type of subset of the *Cartesian product* of A and B. i.e. of the set

$$A \times B = \{(a, b) : a \in A,\ b \in B\}$$

This means that we will always insist that the pairs are *ordered*, i.e. in (a, b) the first coordinate, a, is always an element of A, and the second, b, is an element of B. Moreover, to define a function we shall not allow the set $A \times B$ of *all* ordered pairs itself, but only those subsets which satisfy the following requirement: once we have *fixed* a particular $a \in A$, this should *determine* the second coordinate $b = f(a)$ of the pair (a, b). Thus it will be reasonable to call $b \in B$ the *image* of the element $a \in A$, and write $b = f(a)$, since this image is uniquely determined by a.

Thus: a *function* $f : A \mapsto B$ is a set of ordered pairs (a, b) with $a \in A, b \in B$, such that whenever the pairs (a, b) and (a, b') belong to the set f, then $b = b'$.

Note that this definition does *not* require each $a \in A$ to be the first element of some ordered pair $(a, b) \in f$; we simply require each first element a of such a pair to belong to A, and, similarly, each second element b to belong to B. It seems sensible, however, to give names to the sets of 'all first elements' and 'all second elements' of pairs (a, b); given $f : A \mapsto B$ we define

(i) the *domain of f* as the set

$$\mathcal{D}_f = \{a \in A : \text{there exists } b \in B \text{ such that } (a, b) \in f\}$$

(ii) the *range of f* as the set

$$\mathcal{R}_f = \{b \in B : \text{there exists } a \in A \text{ such that } (a, b) \in f\}$$

More generally, if X is a subset of $\mathcal{D}_f$ we write

$$f(X) = \{f(x) : x \in X\}$$

for the *image* of the set X under the function f. Obviously $f(X)$ is always a subset of the range $\mathcal{R}_f$, and we always have $f(\mathcal{D}_f) = \mathcal{R}_f$.

The *inverse image* of a set Y under the function f is the set of all points x whose images lie in Y, that is

$$f^{-1}(Y) = \{x \in \mathcal{D}_f : f(x) \in Y\}$$

By construction, $f^{-1}(Y)$ is a subset of $\mathcal{D}_f$ and $f^{-1}(\mathcal{R}_f) = \mathcal{D}_f$.

The basic property which ensures that a set of ordered pairs defines a function f is the knowledge that the first coordinate a will always determine the second coordinate $b = f(a)$. A special class of functions of particular interest are functions whose action can be 'undone' unambiguously, i.e. sets of pairs (a,b) where knowledge of the second coordinate b determines the *first* coordinate a. We write the function which 'inverts' $f : A \mapsto B$ in this way as the *inverse function* $f^{-1} : B \mapsto A$; i.e. if f consists of the pairs (a, b) then f^{-1} comprises the corresponding pairs (b, a). It is clear that if we wish f^{-1} to be well-defined as a function, then the original function f must *map distinct points of A to distinct points of B:* that is, if $a \neq a'$ then $f(a) \neq f(a')$. Such a function f is called *one-one* (or *injective*) on its domain.

Now let us bring these rather abstract definitions down to earth by considering *real functions,* i.e. functions for which the sets A, B are both equal to $\mathbb{R}$. Drawing a set of axes in the (x, y)-plane we can see that if $f : \mathbb{R} \mapsto \mathbb{R}$ is given, then the set of points in the plane whose coordinates relative to these axes are the pairs (a, b) comprising f is simply the *graph* of the real function f. Similarly, the domain $\mathcal{D}_f$ is a subset of the x-axis, the range $\mathcal{R}_f$ is a subset of the y-axis, and the condition for f to be a function is simply that no *vertical* line should cut the graph more than once, while f will be one-one precisely when no *horizontal* line cuts the graph more than once.

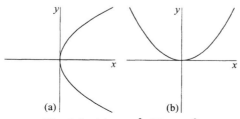

Fig. 5.6 (a) $x = y^2$, (b) $y = x^2$

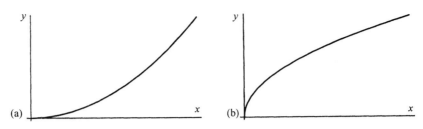

Fig. 5.7 (a) $y = x^2$ $(x \geq 0)$, (b) $y = \sqrt{x}$ $(x \geq 0)$

● *Example 1*

(i) The squaring function $f : x \mapsto x^2$ is well-defined for all real numbers x; but it is not one-one, since $(-1)^2 = 1^2$. On the other hand, if we restrict the function $f(x) = x^2$ to the domain $[0, \infty)$ then the restricted function f is one-one, since if $x \neq x'$ and $x, x' \geq 0$, then $x^2 \neq x'^2$ and we can then discuss $f^{-1} : x \mapsto \sqrt{x}$ as a real function for $x \geq 0$, since only the 'positive square root' comes into play.

(ii) $g(x) = \frac{1}{x}$ is defined for all $x \neq 0$. This function is one-one (and is its own inverse!) but $h = |g|$ is not, since it is the function $h(x) = \frac{1}{|x|}$, which takes the value 1 at both -1 and 1.

(iii) $f(x) = \frac{x-1}{x+1}$ is one-one on the interval $(1, \infty)$. (The graph is drawn below.) Note that the range of f is the interval $(0,1)$ – this can be proved more rigorously later. To find the inverse of f, we need to solve the equation $y = \frac{x-1}{x+1}$ for x. We obtain $f^{-1}(y) = \frac{1+y}{1-y}$ whenever $0 < y < 1$. This example shows that, graphically, finding the inverse corresponds to 'reflecting' the graph of f in the line $y = x$.

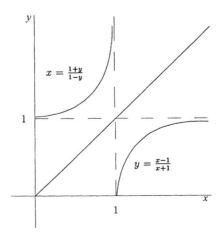

Fig. 5.8   A one-one function and its inverse

Study the examples and numerous sketches given in Chapter 2 of the companion text *Calculus and ODEs* by David Pearson, to develop confidence in handling functions and their graphs.

EXERCISE ON 5.2

1. For each of the following real functions f sketch the graph and identify the domain and range of f. Which of these functions is one-one?

   (i) $f(x) = \sin x + \cos x$ (ii) $f(x) = \sin(\frac{1}{x})$

   (iii) $f(x) = \frac{x^3 - 8}{x^2 - 4}$ (iv) $f(x) = \frac{x}{1 + |x|}$

5.3 Combining real functions

It is natural to regard two real functions f, g as *equal* if their domains $\mathcal{D}_f$ and $\mathcal{D}_g$ coincide and if $f(x) = g(x)$ for all $x \in \mathcal{D}_f = \mathcal{D}_g$. If $\mathcal{D}_f \subset \mathcal{D}_g$ and $f(x) = g(x)$ for all $x \in \mathcal{D}_f$ we call f the *restriction* of g to the set $\mathcal{D}_f$; alternatively we can regard g as an *extension* of f to the set $\mathcal{D}_g$.

Given any two real functions f, g we can form their *sum* $f + g$ by setting $(f + g)(x) = f(x) + g(x)$ and their *product* fg by setting $(fg)(x) = f(x)g(x)$ for all $x \in \mathcal{D}_f \cap \mathcal{D}_g$. Note that we can only define these combinations at points where f and g are *both* defined. More generally, for any real numbers α, β the *linear combination* $(\alpha f + \beta g)$ is well-defined for $x \in \mathcal{D}_f \cap \mathcal{D}_g$ by $(\alpha f + \beta g)(x) = \alpha f(x) + \beta g(x)$. In particular, the *difference* $f - g$ and the *scalar multiples* αf are special linear combinations of f and g.

Slightly more care is needed in defining the *quotient*: first, the *reciprocal* $\frac{1}{g}$ only makes sense when $g(x)$ is defined *and non-zero,* so that the quotient $\frac{f}{g}$ is given by the ratio $(\frac{f}{g})(x) = \frac{f(x)}{g(x)}$ *provided* $x \in \mathcal{D}_f \cap \mathcal{D}_g \cap \{x \in \mathbb{R} : g(x) \neq 0\}$. Note in particular that we can write the quotient as the product of f and $\frac{1}{g}$.

(*Caution: never* confuse the reciprocal $\frac{1}{f}$ with the inverse f^{-1}! They perform quite separate roles, and in general they have quite different domains and ranges.)

Example 2

Any polynomial $P_n(x) = a_0 + a_1 x + a_2 x^2 + \ldots + a_n x^n$ can be built up from the two basic functions $x \mapsto 1$ and $x \mapsto x$, by repeated application of the rules for combining these functions (and multiplying by constants). More generally, a *rational function* is a ratio of two polynomials, such as $f(x) = \frac{3 - 2x + x^3 - x^4}{2 + 3x + x^2}$. To find the domain $\mathcal{D}_f$ we need only consider where the denominator is zero. In this case, $(x^2 + 3x + 2) = (x + 2)(x + 1)$ so the only points to avoid are -2 and -1. Hence $\mathcal{D}_f = \mathbb{R} \setminus \{-2, -1\}$. Finding the range is somewhat trickier and is left for later attention.

A very important operation on functions is that of *composition*: that is, taking a 'function of a function'. We have already seen several examples of this; here is another: compare $f_1 : x \mapsto \sin(x^2)$ and $f_2 : x \mapsto (\sin x)^2$. Note that these are *not* the same thing, since, for example, at $x = \pi$ the value of f_2 is 0, while $f_1(\pi) = \sin(\pi^2)$ is certainly not 0, as the argument π^2 is not an integer multiple of π.

Given two functions f and g we thus need to define both the composition $f \circ g$ and the composition $g \circ f$: this is done simply by setting

$$(f \circ g)(x) = f(g(x)) \qquad (g \circ f)(x) = g(f(x))$$

for all those x for which it makes sense.

For $(f \circ g)(x)$ to be defined we need both $y = g(x)$ and $f(y)$ to be defined, i.e. $x \in \mathcal{D}_g$ and $y = g(x) \in \mathcal{D}_f$, so that for the latter we also need $x \in g^{-1}(\mathcal{D}_f)$. So the domain of $f \circ g$ is the set $\mathcal{D}_g \cap g^{-1}(\mathcal{D}_f)$.

Similarly, $\mathcal{D}_{g \circ f} = \mathcal{D}_f \cap f^{-1}(\mathcal{D}_g)$.

Clearly these two sets need not even coincide, let alone the corresponding function values. For example: let $f(x) = \sqrt{x}$ and $g(x) = \cos(x)$. You should now be able to give precise descriptions of the domains of $f \circ g$ and $g \circ f$.

Intervals and gaps

By far the simplest real functions are those whose domain is an *interval*, that is, a set $I \subset \mathbb{R}$ with the property that if $a, b \in I$ and $a < b$, then every point c satisfying $a \le c \le b$ also belongs to I. These are the subsets of $\mathbb{R}$ which 'inherit' its structure most faithfully. Our intuitive pictures of functions in terms of their graphs can easily deceive us into imagining all functions to have such domains; however, *any* subset of $\mathbb{R}$ can be the domain of some real function. And there are many familiar examples where the domain cannot be an interval, such as $f(x) = \frac{1}{x}$, which is well-defined for $x < 0$ and $x > 0$, but not at $x = 0$. Similarly, the function $g(x) = \sqrt{1 - x^2} + \sqrt{x^2 - 1}$ seems harmless enough, but its domain is the set $\{-1, 1\}$. For such functions the graph clearly *cannot* consist of a 'single unbroken curve'; a gap in the domain will always result in a gap in the graph.

Gaps in the graph can also appear when the domain of the function f is an interval I: for example, the *integer part* function $f(x) = [x]$ assigns to each real number x the largest integer $n \le x$, and therefore has 'jumps' of size 1 at each integer point:

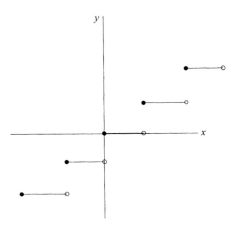

Fig. 5.9 $x \mapsto [x]$

Example 3

As an extreme example, consider the indicator function $1_{\mathbb{Q}}$ defined in Tutorial Problem 5.1. Since every interval contains both rational and irrational numbers, it follows that the graph of $1_{\mathbb{Q}}$ has 'gaps' in *every* interval, however small. (Study the attempted graph once more.)

At the other extreme, an example of a function with a single 'gap' is provided by

$$f(x) = \begin{cases} x & \text{if } x < 3 \\ x^2 - 4 & \text{if } x \geq 3 \end{cases}$$

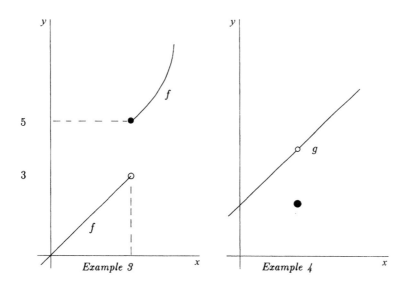

Fig. 5.10 Example 3: jump discontinuity, Example 4: removable discontinuity

The empty circle at the point $(3, 3)$ in the plane denotes that f is *not* defined by the 'endpoint' of that part of the graph: we have $f(3) = 5$ in this example, and this is denoted by the solid circle.

Even though there is just a single gap, we cannot bridge it by changing the value of f at the single point $x = 3$; the parts of the graph will not meet at 3.

Example 4

Now consider the function

$$g(x) = \begin{cases} \frac{x^2 - 4}{x - 2} & \text{if } x \neq 2 \\ 3 & \text{if } x = 2 \end{cases}$$

which similarly has a gap at $x = 2$.

Here we do have a ready-made remedy: to 'fix' the graph so that it remains in one piece, we simply have to define $g(2) = 4$ instead of the given value $g(2) = 3$. To see this, note that for $x \neq 2$, the ratio $\frac{x^2 - 4}{x - 2} = x + 2$, and this 'tends to 4' as x 'approaches 2'. We can make this idea precise in the next section.

1. Suppose that $f(x) = \frac{(1+x)^2 - 1}{x}$ whenever $x \neq 0$. What value must we give this function at 0 if the graph is to have no gaps?

2. Again consider the function $f(x) = \frac{x^3 - 8}{x^2 - 4}$ considered in Exercise on 5.2. Show that one of the 'gaps' in the graph can be corrected by giving the function an appropriate value at one point, while the other cannot.

5.4 Limits of real functions – what Cauchy meant!

The definitions of *limit* and *continuity* really only became well established following the publication of *Augustin-Louis Cauchy's* famous treatise *Cours d'Analyse* in 1821. Cauchy tried to base his definitions and theorems on as few fundamental concepts as he could, and insisted that he would dispense with pictures in his proofs. Thus the concept of *variable* became the basic tool, and a *function* of a variable described a relationship between an 'independent' variable x and a 'dependent' one, $y = f(x)$. Without pursuing this train of thought too far (since Cauchy also stated that the 'calculus of infinitesimals' was 'indispensible' for his analysis) it soon becomes clear that Cauchy's variables take on values in succession, i.e. *sequentially*, and that his definition of the *limit of a function at a point* makes essential use of the concept of *limit of a sequence*. Without fully repeating his definition here, we can take up this idea and cast it in our modern terminology:

Fig. 5.11 Augustin-Louis Cauchy (1789–1857)

Augustin-Louis Cauchy – the 'Father of Modern Analysis'(?) – whose Cours d'Analyse *dominated the development of the subject between 1820 and 1850. Yet Cauchy (like many of his successors) could not quite tear himself away from the attractions of infinitesimals, despite the inherent difficulty in making sense of these elusive quantities.*

• Definition 1

Let f be a real function, let $a \in \mathbb{R}$ be given, and suppose that there is at least one sequence $(x_n) \subset \mathcal{D}_f$ converging to a and such that $x_n \neq a$ for all $n \geq 1$. We say that the real number L is the *limit* of $f(x)$ as x approaches a if, for *every* sequence (x_n) in $\mathcal{D}_f$ such that $x_n \neq a$ and $x_n \to a$ as $n \to \infty$, we also have $f(x_n) \to L$ as $n \to \infty$. We write this as $\lim_{x \to a} f(x) = L$.

Note that in the example at the end of Section 5.3 we have

$$\lim_{x \to a} \frac{x^2 - 4}{x - 2} = 4$$

since if $x_n \neq 2$, we have $\frac{x_n^2 - 4}{x_n - 2} = x_n + 2$, and if $x_n \to 2$, $x_n + 2 \to 4$ as $n \to \infty$. Thus using $\lim_{x \to 2} g(x) = 4$ as the value to give the function g defined for $x \neq 2$ by $g(x) = \frac{x^2 - 4}{x - 2}$ will ensure that the graph of the extended function 'remains in one piece'.

Note that we did not need to have a value for $f(a)$ in order to define $L = \lim_{x \to a} f(x)$. Thus the point a need not belong to $\mathcal{D}_f$ for the limit of the function to exist at this point. However, it will be useful to know what happens to f in a *neighbourhood* of a; by this we simply mean a set of the form $N(a, \delta) = \{x : |x - a| < \delta\}$ for some $\delta > 0$. In fact, the definition and properties of limits become a little tricky unless we assume that f is defined *throughout* some neighbourhood of the point a in question (except, that is, possibly at a itself: we shall call the set $N'(a, \delta) = \{x : 0 < |x - a| < \delta\} = N(a, \delta) \setminus \{a\}$ a *punctured* (or *deleted*) *neighbourhood of a*. This deals with most eventualities, and from now on we shall usually make this simplifying assumption in our examples.

The first important exception comes when a is 'at the edge' of $\mathcal{D}_f$, as for example, when $f(x) = \sqrt{x}$ (or $g(x) = \frac{1}{\sqrt{x}}$) and $a = 0$. Note that in these examples a is in the domain of f but not in the domain of g.

To deal with both cases we introduce *one-sided limits*:

• Definition 2

Let $a \in \mathbb{R}$ be given and suppose that the domain of $f : \mathbb{R} \mapsto \mathbb{R}$ contains the open interval $(a, a + \delta)$ for some $\delta > 0$. We say that $L+$ is the *right-hand limit of f as x approaches a* if, given any sequence (x_n) in $\mathcal{D}_f$ with $x_n > a$ for all $n \geq 1$, and $\lim_{n \to \infty} x_n = a$, we have $\lim_{n \to \infty} f(x_n) = L+$. We write this as $L+ = \lim_{x \downarrow a} f(x)$.

Similarly, for any f such that $\mathcal{D}_f$ contains some open interval $(a - \delta, a)$, we define $L-$ as the *left-hand limit of f as x approaches a*, if, given any sequence (x_n) in $\mathcal{D}_f$ with $x_n < a$ for $n \geq 1$ and $\lim_{n \to \infty} x_n = a$ we have $\lim_{n \to \infty} f(x_n) = L-$. Now we write $L- = \lim_{x \uparrow a} f(x)$.

TUTORIAL PROBLEM 5.3

Show that $\lim_{x \to a} f(x) = L$ if and only if L is the common value of the left- and right-hand limits of f at a.

Example 5

For examples where $L+$ and $L-$ are different, consider the following:

(i) $f(x) = \begin{cases} 1-x & \text{if } x<1 \\ x^2 & \text{if } x\geq 1 \end{cases}$

(ii) $g(x) = [x]$ (recall that $[x]$ denotes the largest integer $n \leq x$).

In (i) $\lim_{x\uparrow 1} f(x) = \lim_{x\uparrow 1}(1 - x) = 0$, while $\lim_{x\downarrow 1} f(x) = \lim_{x\downarrow 1} x^2 = 1$. In (ii) the left-hand limit of g at the integer n is $(n - 1)$, since g is constant at that value on the interval $[n - 1, n)$, while the right-hand limit of g at n is n.

A second case where we need to amend our idea of 'neighbourhood' somewhat arises when we consider the limit of a function as x 'tends to infinity', as for example with $f(x) = \frac{1}{x}$: it is clear that we need $\lim_{x\to\infty} f(x) = 0$, since for any sequence, if $x_n \to \infty$ then $\frac{1}{x_n}$ is arbitrarily close to 0 if we take n sufficiently large. (Recall from Chapter 2 that $x_n \to \infty$ means that, given $K > 0$, we can find a natural number N such that $x_n > K$ whenever $n > N$.) Thus our definition must read:

Suppose that the domain of the real function f contains the interval (K, ∞) for some $K > 0$. Then we define $L = \lim_{x\to\infty} f(x)$ if $f(x_n) \to L$ whenever $x_n \to \infty$. We say that L is the *limit at infinity* of f.

A similar definition (which you should now write out in full!) applies to $\lim_{x\to-\infty} f(x)$. Note that in each case an interval of the form (K, ∞) or $(-\infty, -K)$ plays the role of 'neighbourhood' of the limit point.

Example 6

Let $f(x) = \frac{1}{1+x^2}$, which is well-defined for all real x. It should be clear that $\lim_{x\to\infty} f(x) = 0 = \lim_{x\to-\infty} f(x)$.

Similarly, $g(x) = \cos(\frac{1}{x})$ is well-behaved as $x \to \infty$ or $x \to -\infty$, and we shall see later that $\lim_{x\to\infty} \cos(\frac{1}{x}) = 1$. However, things are much wilder as x approaches 0: in fact $\cos(\frac{1}{x})$ has no limit (not even a one-sided one) at 0. To see this, consider the sequence (x_n) with $x_n = \frac{1}{\pi n}$. On the one hand (x_n) converges to 0 as $n \to \infty$, but on the other, $\cos(\frac{1}{x_n}) = \cos(\pi n) = (-1)^n$, and this sequence has no limit. The same applies with $y_n = -\frac{1}{\pi n}$, so neither one-sided limit can exist at 0.

However, the function $h(x) = x\cos(\frac{1}{x})$ has $\lim_{x\to 0} h(x) = 0$, since $|x\cos(\frac{1}{x})| \leq |x|$ for all x, so that for any sequence $x_n \to 0$ we have $h(x_n) \to 0$.

We can translate our 'sequential' definition of limits of functions to one which uses the new concept of the 'punctured neighbourhood' $N'(a, \delta)$ of a given point $a \in \mathbb{R}$: when we consider the graph of the function f, the number $L = \lim_{x\to a} f(x)$ is the unique number for which we can ensure that $f(x)$ approximates L *as closely as we please* as long as we take $x \neq a$ *sufficiently close to a.* In other words:

● *Theorem I* ─────────────────────────────────

Let f be a real function, $a \in \mathbb{R}$. Then $L = \lim_{x \to a} f(x)$ if and only if L satisfies the following condition:

> ($\varepsilon - \delta$ condition for limits): given $\varepsilon > 0$ there is a $\delta > 0$ such that $|f(x) - L| < \varepsilon$ whenever $0 < |x - a| < \delta$.

PROOF

If the $\varepsilon - \delta$ condition holds, and (x_n) is a sequence in $\mathcal{D}_f$ with $x_n \neq a$ and $x_n \to a$ as $n \to \infty$, then we can find $N \in \mathbb{N}$ such that $|x_n - a| < \delta$ whenever $n \geq N$. By hypothesis this ensures that $|f(x_n) - L| < \varepsilon$ for $n \geq N$. But this means that the sequence $(f(x_n))_n$ converges to L, as required.

On the other hand, suppose the $\varepsilon - \delta$ condition fails to hold. This means that there *exists* an $\varepsilon > 0$ such that *for each* $\delta > 0$ we can find some $x \in \mathcal{D}_f$ which satisfies $0 < |x - a| < \delta$, but $|f(x) - L| \geq \varepsilon$. For this *fixed* $\varepsilon > 0$ we can successively choose $\delta = 1, \frac{1}{2}, \frac{1}{3}, ..., \frac{1}{n}, ...$ and find a sequence (x_n) as follows:

$x_1 \in \mathcal{D}_f$ satisfies $0 < |x_1 - a| < 1$ but $|f(x_1) - L| \geq \varepsilon$

$x_2 \in \mathcal{D}_f$ satisfies $0 < |x_2 - a| < \frac{1}{2}$ but $|f(x_2) - L| \geq \varepsilon$

$x_3 \in \mathcal{D}_f$ satisfies $0 < |x_3 - a| < \frac{1}{3}$ but $|f(x_3) - L| \geq \varepsilon$

....

$x_n \in \mathcal{D}_f$ satisfies $0 < |x_n - a| < \frac{1}{n}$ but $|f(x_n) - L| \geq \varepsilon$

....

The sequence (x_n) clearly converges to a, but since the distances between $f(x_n)$ and L are always at least ε, we cannot have $f(x_n) \to L$. Thus $L \neq \lim_{x \to a} f(x)$. Hence if $L = \lim_{x \to a} f(x)$ the $\varepsilon - \delta$ condition must hold.

We have shown that the two conditions are *equivalent*. Therefore either of them could be used to define the limit of $f(x)$ as x tends to a. Having proved their equivalence, we are now free to use either form in dealing with examples.

TUTORIAL PROBLEM 5.4

(i) Again it should be easy to see what is required for a 'one-sided' version of the $\varepsilon - \delta$ condition to hold – give a careful proof of the following 'right-hand version' and write down its 'left-hand' counterpart:
$L+ = \lim_{x \downarrow a} f(x)$ if and only if: given $\varepsilon > 0$ there exists $\delta > 0$ such that $|f(x) - L+| < \varepsilon$ whenever $a < x < a + \delta$.

(ii) Show that the '$\varepsilon - \delta$' version of $L = \lim_{x \to \infty} f(x)$ is: for given $\varepsilon > 0$ there exists $K > 0$ such that $|f(x) - L| < \varepsilon$ whenever $x > K$.

(iii) Things become rather more subtle if it is possible that there is *no* sequence $(x_n) \in \mathcal{D}_f$ with $x_n \neq a$ but $x_n \to a$. To avoid this possibility we assumed in our definition of $\lim_{x \to a} f(x)$ that there is at least one sequence (x_n) in $\mathcal{D}_f$ converging to a for which none of the x_n equal a; i.e. that a is at least an *accumulation point* of the domain of f (see Chapter 3). Discuss how this and the neighbourhood concept fit together, and what can happen if a fails to satisfy these conditions.

We shall return to these questions in the context of continuous functions.

● *Example 7*

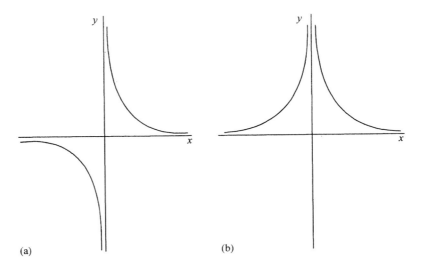

(a)

(b)

Fig. 5.12 (a) $y = \frac{1}{x}$, (b) $y = \frac{1}{|x|}$

The graph of $f(x) = \frac{1}{x}$ shows that f cannot have a (finite) limit at $x = 0$, since the sequences $x_n = \frac{1}{n}$ and $x'_n = -\frac{1}{n}$ both converge to 0, but $f(x_n) = n$ diverges to $+\infty$ and $f(x'_n) = -n$ to $-\infty$. However, comparing this with $g(x) = \frac{1}{|x|}$ for example, we are led to defining limits for unbounded functions. Here it is convenient to use neighbourhoods instead of sequences:

$\lim_{x \to a} f(x) = +\infty$ if for any $K > 0$ there exists $\delta > 0$ such that for all $x \in \mathcal{D}_f \cap N'(a, \delta)$ we have $f(x) > K$.

$\lim_{x \downarrow a} f(x) = +\infty$ if for any $K > 0$ there exists $\delta > 0$ such that for all $x \in \mathcal{D}_f$ with $a < x < a + \delta$ we have $f(x) > K$.

The corresponding definitions for $-\infty$ are clear. Hence $\lim_{x \to 0} (\frac{1}{|x|}) = \infty$, while for $f(x) = \frac{1}{x}$ the limit at 0 fails to exist, since the right-hand limit is $+\infty$ and the left-hand limit is $-\infty$.

The Algebra of Limits (Theorem 1 in Chapter 2) guarantees that for two-sided limits we can combine limits in the expected fashion:

● *Theorem 2*

Suppose that f, g are real functions whose domains contain a punctured neighbourhood of $a \in \mathbb{R}$. Let α, β be real numbers, and let $L = \lim_{x \to a} f(x)$, $M = \lim_{x \to a} g(x)$. Then:

(i) $\lim_{x \to a}(\alpha f + \beta g)(x) = \alpha L + \beta M$

(ii) $\lim_{x \to a}(f \cdot g)(x) = LM$

(iii) $\lim_{x \to a}(\frac{f}{g})(x) = \frac{L}{M}$ when $M \neq 0$.

PROOF

This follows at once from Theorem 1 in Chapter 2, since if a sequence $(x_n) \in \mathcal{D}_f \cap \mathcal{D}_g$ converges to a, we have $f(x_n) \to L$ and $g(x_n) \to M$, so that $(\alpha f + \beta g)(x_n) \to \alpha L + \beta M$ and $(fg)(x_n) \to LM$, while if $M \neq 0$, we also have $(\frac{f}{g})(x_n) \to \frac{L}{M}$ as $n \to \infty$.

TUTORIAL PROBLEM 5.5

Discuss which 'one-sided' versions of the above result remain true.

EXERCISES ON 5.4

1. Let $f(x) = \frac{3x^2 - 2x + 1}{x^3 + 4x + 2}$. Find $\lim_{x \to 2} f(x)$.

2. Suppose that $f(y) = \begin{cases} 2 & \text{if } y = 1 \\ 0 & \text{if } y \neq 1 \end{cases}$ and that $g(x) = 1$ for all $x \in \mathbb{R}$. Show that $f(y) \to 0$ as $y \to 1$ and $g(x) \to 1$ as $x \to a$ for any $a \in \mathbb{R}$, but that $f \circ g(x)$ does not have limit 0 as $x \to a$. What does this tell us about the composition of limits?

Summary

This chapter has been devoted to a discussion of the modern concept of *real function*, which has been defined using *sets* as the basic undefined notion. The long and convoluted history of functions throughout the development of the Calculus was outlined, and the basic concept of *limit* of a function at a point was introduced following Cauchy's original sequential definition. This had the advantage of enabling us to use the Algebra of Limits results proved for sequences in order to show how limits are combined, and also gave insight into the use of one-sided limits.

We also stressed the importance of the idea of *neighbourhood* of a point: in what follows limits will only be discussed when the function concerned is defined throughout a punctured neighbourhood of the point in question.

FURTHER EXERCISES

1. The *sign* function is defined as

$$sgn(x) = \begin{cases} 1 & \text{if } x > 0 \\ 0 & \text{if } x = 0 \\ -1 & \text{if } x < 0 \end{cases}$$

   Show that $\lim_{x \to 0} sgn(x)$ does not exist, but that both one-sided limits exist at 0. Verify that $sgn(x) = \frac{|x|}{x}$ for $x \neq 0$. Why does this suggest that we can use *sgn* as the 'derivative' of the modulus function?

2. Use the $\varepsilon - \delta$ condition for limits to show that $\lim_{x \to a} \sqrt{x} = \sqrt{a}$ for all $a > 0$.

3. Show that $\lim_{x \to 1} \frac{x^2+x+1}{x^2+2x+1} = \frac{3}{4}$ in two ways: (i) directly from the $\varepsilon - \delta$ condition, and (ii) by using Theorem 2.

4. Use the corresponding results for sequences to prove the following facts relating limits to inequalities:

   (i) If $f(x) \leq g(x)$ for all $x \neq a$ and if $\lim_{x \to a} f(x) = L$ and $\lim_{x \to a} g(x) = M$, then $L \leq M$.

   (ii) *The sandwich principle*:
   If $f \leq g \leq h$ throughout some punctured neighbourhood of a, and $\lim_{x \to a} f(x) = L = \lim_{x \to a} h(x)$, then also $\lim_{x \to a} g(x) = L$.

6 • Continuous Functions

Intuitively, as well as historically, a *continuous function f* is one whose graph contains no jumps, and can therefore been drawn 'without lifting pencil from paper'. This isn't such a bad idea, as many of our examples will show. However, as we have seen, mathematicians have not always agreed on what should be regarded as a *function*, and our simple intuitive pictures are not always a reliable guide to the definitions we now employ. These problems provoked much discussion in the eighteenth and nineteenth centuries, when many mathematicians, still regarding real functions primarily as curves in the plane, argued that to be a 'genuinely' continuous function, *f* had to be defined by a single formula, and introduced distinctions between 'continuity' of functions and 'contiguity' of their graphs. These ideas lost ground with the introduction of such natural ideas as describing the modulus as the function:

$$f(x) = |x| = \begin{cases} x & \text{if } x \geq 0 \\ -x & \text{if } x < 0 \end{cases}$$

and with various functions such as the indicator function $\mathbf{1}_\mathbb{Q}$ of the set $\mathbb{Q}$ of rational numbers. The latter function 'jumps' at every point of $\mathbb{R}$, and will have to be described carefully without much help from geometric intuition, since drawing the graph accurately is quite impossible.

We will use the machinery of limits developed in the earlier chapters to provide a unified approach to these problems, and to describe more precisely the scope of the 'naive' idea of continuity mentioned above.

6.1 Limits that fit

The basic idea is very simple, but important enough to state it as:

• Definition I

Let f be a real function and let $a \in \mathcal{D}_f$ be given. We say that f is *continuous at a* if $\lim_{x \to a} f(x)$ exists and equals $f(a)$.

If f is continuous at every point of a set A we say that f is continuous *on A*. A *continuous function* is one which is continuous on its domain.

Thus f is continuous if all its limits exist *and* 'fit the graph' of f correctly, i.e. if $L = \lim_{x \to a} f(x)$ then $L = f(a)$. However, this definition turns out to be somewhat restrictive: recall that we only *defined* $\lim_{x \to a} f(x)$ when there is at least one non-constant sequence in $\mathcal{D}_f$ converging to a. Taking our cue from the Definition 2 in Chapter 3, we now define a point a as an *accumulation point* of the set $A \subset \mathbb{R}$ if there is a sequence $(x_n) \subset A$ with $x_n \to a$ as $n \to \infty$, but $x_n \neq a$ for each $n \geq 1$. In other words, every neighbourhood of a contains points of A other than a itself.

Thus our definition of $\lim_{x \to a} f(x)$ makes sense only at accumulation points of $\mathcal{D}_f$, even though the point a itself does not need to belong to $\mathcal{D}_f$. (In our examples in Chapter 5 we even assumed that $\mathcal{D}_f$ should contain a (punctured) neighbourhood of a.)

In order to have a general definition of continuity at points of $\mathcal{D}_f$, we need to include the remaining points of $\mathcal{D}_f$: we shall call a an *isolated point* of $\mathcal{D}_f$ if it is not an accumulation point, and we shall decree that every real function is *always* continuous at all isolated points of its domain.

Of course, if f is defined throughout some neighbourhood of a – as will invariably be true in the examples we consider in this book – then a is not an isolated point of $\mathcal{D}_f$! Thus the above extension of Definition 1 serves simply to avoid unpleasant counter-examples to some of our more desirable theorems.

We shall proceed to use the results of Chapter 5 directly:

● *Theorem 1* ——————————————————————————

Suppose f and g are continuous at the point $a \in \mathbb{R}$, and $\alpha, \beta \in \mathbb{R}$ are given. Then:

  (i) $\alpha f + \beta g$ is continuous at a;
 (ii) $f . g$ is continuous at a;
(iii) if $g(a) \neq 0$ then $\frac{f}{g}$ is continuous at a.

PROOF
This follows at once from Theorem 2 in Chapter 5 if we set $L = f(a)$ and $M = g(a)$.

Since the limits now fit correctly, we can also show that continuity is preserved under composition of functions:

● *Theorem 2* ——————————————————————————

If f is continuous at a and g is continuous at $b = f(a)$, then the composition $g \circ f$ is continuous at a.

PROOF
For $g \circ f$ to be defined at x we need x to be in the domain of f and $f(x)$ to be in the domain of g, so that $(g \circ f)(x) = g(f(x))$ makes sense. Take a sequence (x_n) with these properties, and such that $x_n \to a$ as $n \to \infty$. Let $y_n = f(x_n)$. Then $y_n \to f(a) = b$ since f is continuous at a. Thus we have a sequence (y_n) in the domain of g such that $y_n \to b \in \mathcal{D}_g$ as $n \to \infty$. As g is continuous at b, we conclude that $(g \circ f)(x_n) = g(y_n) \to g(b) = (g \circ f)(a)$ as $n \to \infty$.

● *Example 1*

We start with 'trivial' cases: the constant function $f : x \mapsto 1$ has domain $\mathbb{R}$ and is certainly continuous, since if $x_n \to a$, then $\{f(x_n) : n \geq 1\}$ is the constant sequence $\{1, 1, 1, \ldots\}$, hence certainly converges to 1. Similarly, $f : x \mapsto x$ has domain $\mathbb{R}$ and since $f(x) = x$ for all $x \in \mathbb{R}$, it follows that $f(x_n) = x_n$ for any sequence. So the statements '$x_n \to a$' and '$f(x_n) \to f(a)$' are *the same* in this case. Hence f is also continuous.

But now we can use Theorem 1 repeatedly to show that every *polynomial* $P(x) = a_0 + a_1x + a_2x^2 + \ldots + a_nx^n$ *must be continuous at every point of* $\mathbb{R}$; similarly, we see that any *rational function* (recall that this means a ratio of polynomials $\frac{P}{Q}$) is continuous at all points of its domain (which is when $Q(x) \neq 0$). Thus, for example,

$$x \longmapsto \frac{2x^9 - 3.5x^6 + 4x^4 - 2x^3 + 11}{x^3 - 3x^2 + 2x - 7}$$

is continuous at each point where the denominator $\neq 0$. Moreover, using Theorem 2 and *assuming* for the moment that the function sin is continuous, we can even see immediately that

$$x \longmapsto \left\{ \sin\left(\frac{2x^9 - 3.5x^6 + 4x^4 - 2x^3 + 11}{x^3 - 3x^2 + 2x - 7}\right) \right\}^3$$

is continuous whenever it is defined!

⊕ *Example 2*

To show directly that the square root function is continuous at each point $a > 0$, we have to work a little harder (*see also* Further Exercise 2 at the end of Chapter 5): suppose $x_n \to a$ as $n \to \infty$, where each $x_n > 0$, and $a > 0$. Can we say that $\sqrt{x_n} \to \sqrt{a}$? To see that we *can*, we again employ the 'difference of two squares':

$$x - a = (\sqrt{x} - \sqrt{a})(\sqrt{x} + \sqrt{a})$$

so that for all $n \geq 1$,

$$|\sqrt{x_n} - \sqrt{a}| = \frac{|x_n - a|}{\sqrt{x_n} + \sqrt{a}} < \frac{1}{\sqrt{a}}|x_n - a| \to 0$$

as $n \to \infty$. Hence $f : x \longmapsto \sqrt{x}$ is continuous at each $a > 0$.

We can reformulate the *neighbourhood condition* of Theorem 1 in Chapter 5 to give us an alternative approach to continuity at a point:

● *Theorem 3* ───────────────────

Let f be a real function, $a \in \mathcal{D}_f$. Then f is continuous at a if and only if it satisfies the following condition:

> ($\varepsilon - \delta$ condition for continuity): given $\varepsilon > 0$ there is a $\delta > 0$ such that $|f(x) - f(a)| < \varepsilon$ whenever $|x - a| < \delta$.

This follows immediately from Theorem 1 in Chapter 5: the only change is that we now demand that f is defined at a and $L = f(a)$. We can dispense with the condition $0 < |x - a|$, since omitting it only requires us to check that $|f(x) - f(a)| < \varepsilon$ also holds for $x = a$, which is trivially true. (Note that for an isolated point $a \in \mathcal{D}_f$ there is also nothing to prove, since there will be some $N(a, \delta)$ containing no points of $\mathcal{D}_f$ other than a!)

Using the $\varepsilon - \delta$ condition to prove continuity of $x \mapsto \sqrt{x}$ at $a \neq 0$ requires that, for given $\varepsilon > 0$ we should find $\delta > 0$ such that $|x - a| < \delta$ implies $|\sqrt{x} - \sqrt{a}| < \varepsilon$. Show that setting $\delta = \sqrt{a}.\varepsilon$ will ensure that δ satisfies our conditions. Note that this illustrates a property of the square root function which is already apparent from its graph: when a is near 0 the gradient is steep and δ may have to be chosen to be much smaller than the given 'error bound' ε, while for large a, the choice of a fairly 'large' δ may suffice. Illustrate this by choosing a number of different values for a and computing δ.

Finally, we reformulate the $\varepsilon - \delta$ condition entirely in terms of neighbourhoods; in this form it will generalize more easily to the abstract situations considered in more advanced treatments of analysis. Note that the inequality $|f(x) - f(a)| < \varepsilon$ means that the value (i.e. the *real number*) $f(x)$ belongs to the ε−neighbourhood of $f(a)$, that is, $f(x) \in N(f(a), \varepsilon)$. Similarly the inequality $|x - a| < \delta$ just says that $x \in N(a, \delta)$. Thus the demand that f is continuous at a requires that for *every* $x \in N(a, \delta)$ we should have $f(x) \in N(f(a), \varepsilon)$. Hence the *image* under f of the neighbourhood $N(a, \delta)$ must lie wholly inside $N(f(a), \varepsilon)$, i.e. we have formulated the

Neighbourhood condition for continuity of f at a: for each $\varepsilon > 0$ there exists $\delta > 0$ such that $f(N(a, \delta)) \subset N(f(a), \varepsilon)$.

This has some useful consequences:

• Proposition 1

Let $f : \mathbb{R} \mapsto \mathbb{R}$ be continuous at a. Then:

(i) there is a neighbourhood $N(a, \delta)$ on which f is bounded;
(ii) if $f(a) > 0$, there is a neighbourhood $N(a, \delta)$ and a number $c > 0$ such that $f(x) > c$ for all $x \in N(a, \delta)$.

PROOF
(i) For any $\varepsilon > 0$ (for example we could take $\varepsilon = 1$!) we can find $\delta > 0$ such that $f(N(a, \delta)) \subset N(f(a), \varepsilon)$, hence for x throughout $N(a, \delta)$ the values $f(x)$ lie between $f(a) - \varepsilon$ and $f(a) + \varepsilon$.
(ii) Take $c = \frac{1}{2}f(a)$ and choose $\delta > 0$ so that $f(N(a, \delta) \subset N(f(a), c)$. Then $f(x) > c$ for all $x \in N(a, \delta)$.

⊕ Example 3

We illustrate these ideas for the function f defined by:

$f(x) = \frac{1}{x}$ for $x \neq 0$. For $a > 0$ let $\delta = \frac{a}{2}$, then $f(x) \in (\frac{2}{3a}, \frac{2}{a})$ whenever $x \in N(a, \delta)$, and so f is both bounded above by $\frac{2}{a}$ and strictly greater than $\frac{2}{3a} > 0$ throughout this neighbourhood of a.

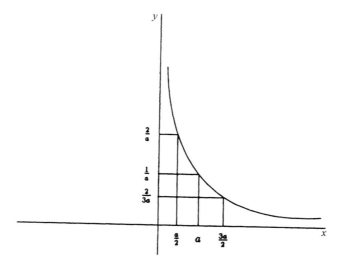

Fig. 6.1. $y = \frac{1}{x}$ (neighbourhoods)

Example 4

Some other examples we use frequently are:

(i) the *modulus function*: $x \mapsto |x|$, which is continuous, since we know that $||x_n| - |a|| \le |x_n - a|$.

(ii) the *integer part function*: $x \mapsto [x]$, which is *discontinuous* at each integer k: to 'see' this we need only draw the graph, which has constant height $k - 1$ on the interval $[k - 1, k)$, but jumps to height k at $x = k$. A *proof*, using Definition 1, can be given as follows: take a sequence (x_n) with

$$x_n = \begin{cases} k - \frac{1}{n} & \text{if } n \text{ is odd} \\ k + \frac{1}{n} & \text{if } n \text{ is even} \end{cases}$$

Then $x_n \to k$ as $n \to \infty$, but $f(x_n) = [x_n] = \begin{cases} k-1 & \text{for odd } n \\ k & \text{for even } n \end{cases}$. Hence $(f(x_n))_n$ does not converge, so f is not continuous at k.

This function has an interesting 'cousin': let $g(x) = x - [x]$, then the horizontal lines turn through $45°$ and the values repeat in each interval of length 1, so that the function has *period 1*. It is continuous on each open interval $(n, n + 1)$, but discontinuous at integer points.

(iii) the *indicator function of the rationals*, i.e. $f = \mathbf{1}_{\mathbb{Q}}$ (defined earlier) is *discontinuous at every point* of $\mathbb{R}$. To see this for $a \in \mathbb{Q}$, simply choose a sequence (a_n) of irrationals converging to a, so that $f(a_n) = 0$ for all n, while $f(a) = 1$. For $a \in \mathbb{R} \setminus \mathbb{Q}$ choose rationals (a_n) converging to a. Now $f(a) = 0$ while each $f(a_n) = 1$. So in each case $f(a_n)$ fails to converge to a.

TUTORIAL PROBLEM 6.2

Discuss the graphs of the above functions.

1. Let $f : \mathbb{R} \mapsto \mathbb{R}$ be defined by $f(x) = \begin{cases} x & \text{if } x \in \mathbb{Q} \\ 1-x & \text{if } x \notin \mathbb{Q} \end{cases}$. Try to 'sketch the graph' of f. Show that f is continuous only at $x = \frac{1}{2}$.

2. Suppose that f is continuous on $\mathbb{R}$ and $f(x) = 0$ at each rational x. Show that $f(x) = 0$ for every real number x. (*Hint*: recall that any interval contains both rationals and irrationals.)

3. What value must we give $x \mapsto x^2 \cos(\frac{1}{x^2})$ at $x = 0$ in order to obtain a continuous extension of this function to $\mathbb{R}$?

6.2 Limits that do not fit: types of discontinuity

The language of limits allows us to classify a number of different reasons why a function can fail to be continuous at a given point in its domain. First we extend the link between continuity and limits to the one-sided versions of the definition:

$f : \mathbb{R} \mapsto \mathbb{R}$ is *left-continuous at* $a \in \mathcal{D}_f$ if $\lim_{x \uparrow a} f(x) = f(a)$ and *right-continuous at* $a \in \mathcal{D}_f$ if $\lim_{x \downarrow a} f(x) = f(a)$. Thus f is continuous at $a \in \mathcal{D}_f$ if and only if it is both left- and right-continuous at a.

● *Example 5*

$x \mapsto \sqrt{x}$ is right-continuous at 0, since $x_n \downarrow 0$ implies $\sqrt{x_n} \downarrow 0$ as $n \to \infty$. (Recall that we proved $x_n^p \downarrow 0$ for all $p > 0$ in Example 10 in Chapter 2!)

The distinction between two-sided and one-sided continuity leads to the first class of discontinuities below.

● *Example 6—Types of discontinuity*

(i) f has a *jump discontinuity at* $a \in \mathcal{D}_f$ if both one-sided limits exist at a, but are not equal. This can happen in one of two ways: the jump can be *bounded*, as is the case with the integer-part function $f(x) = [x]$, which is right-continuous at integer points, since $\lim_{x \downarrow n} f(x) = n = f(n)$, but not left-continuous there, since $\lim_{x \uparrow n} f(x) = n - 1 < f(n)$ for each integer n.

The second possibility is that the jump is *unbounded*: an example is given by

$$g(x) = \begin{cases} \frac{1}{x} & \text{if } x > 0 \\ 0 & \text{if } x \leq 0 \end{cases}$$

at $a = 0$, since $\lim_{x \downarrow 0} g(x) = +\infty$, while $g(0) = 0$. We also say that g has an *infinite singularity* at 0, since no attempt to redefine g at 0 will make it right-continuous there.

(ii) Even if $L = \lim_{x \to a} f(x)$ exists there is no guarantee that f is continuous at a, since we need, in addition, that $L = f(a)$. So if f is given the 'wrong' value at a it can be discontinuous there: let

$$f(x) = \begin{cases} 1 - x & \text{if } x < 1 \\ 1 & \text{if } x = 1 \\ x - 1 & \text{if } x > 1 \end{cases}$$

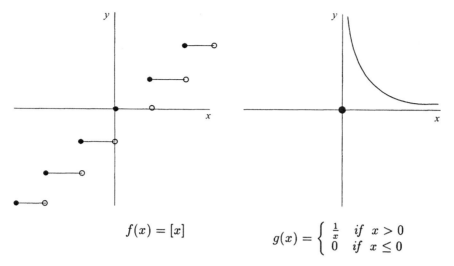

$$f(x) = [x]$$

$$g(x) = \begin{cases} \frac{1}{x} & \text{if } x > 0 \\ 0 & \text{if } x \le 0 \end{cases}$$

Fig. 6.2. Bounded and unbounded jumps

Here the discontinuity at $x = 1$ can be removed by defining $f(1)$ to be the value of $\lim_{x\to 1} f(x) = 0$ instead. Thus this example illustrates a *removable discontinuity*: all that is needed to make f continuous at a is to redefine $f(a)$ as $\lim_{x\to a} f(x)$.

(iii) The 'wildest' situation is that neither the left- nor the right-hand limits of f exist at a: an example of this is $f(x) = \cos(\frac{1}{x})$, as we saw at the end of Chapter 5. Here the reason for the lack of continuity – which occurs at $a = 0$ whatever value we choose for $f(0)$ – is the rapid oscillation of the function near the point; hence such discontinuities are called *oscillatory*. In this example the oscillations are bounded, since $\sin(x)$ always takes values in $[-1, 1]$. However, the function

$$h(x) = \begin{cases} 0 & \text{if } x = 0 \\ \frac{1}{x}\cos(\frac{1}{x}) & \text{otherwise} \end{cases}$$

has unbounded oscillations near 0, since its values in the interval $[\frac{1}{(n+2)\pi}, \frac{1}{n\pi}]$ range throughout the interval $[-(n-1)\pi, (n+2)\pi]$ for any $n \in \mathbb{N}$.

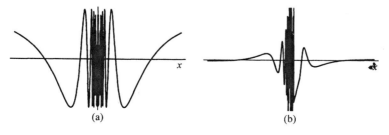

(a) (b)

Fig. 6.3. (a) $\left(\frac{1}{x}\right)$, (b) $\frac{1}{x}\cos\left(\frac{1}{x}\right)$

● *Example 7*

We can now consider *Dirichlet's function* (*see* Tutorial Problem 5.1) in more detail; this will again illustrate the usefulness of Theorem 3.

Define the function $f : [0, 1) \mapsto \mathbb{R}$ as follows:

$$f(x) = \begin{cases} \frac{1}{n} & \text{if } x = \frac{m}{n} \in \mathbb{Q} \\ 0 & \text{if } x \notin \mathbb{Q} \end{cases}$$

So $f(\frac{1}{2}) = \frac{1}{2}$, $f(\frac{1}{3}) = f(\frac{2}{3}) = \frac{1}{3}$, $f(\frac{1}{4}) = f(\frac{3}{4}) = \frac{1}{4}$, etc. Drawing a 'graph' of f is impossible, except for a few rational points, and there f certainly does not look continuous!

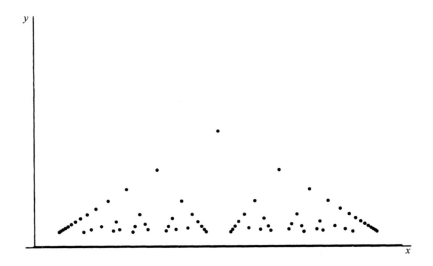

Fig. 6.4. Dirichlet's function

But is f continuous *anywhere*, given that for any $x \in \mathbb{R}$ we can find a sequence of rationals converging to x? Surprisingly, we can show that:

f is continuous at each irrational and discontinuous at each rational point.

Fix an $n \in \mathbb{N}$. The set of points x at which $f(x) \geq \frac{1}{n}$ consists of all rationals whose denominator is no greater than n. There are no more than $\frac{1}{2}n(n-1)$ such points, since when $x = \frac{m}{n} \in (0, 1)$, we have $m < n$. Given $\varepsilon > 0$, take $n > \frac{1}{\varepsilon}$. Fix an $a \in (0, 1)$, and choose $\delta > 0$ small enough to ensure that none of the above points (unless a is itself one of them!) lies in the interval $I_\delta = (a - \delta, a + \delta)$. This can be done, since a lies a positive distance from each of the other points. Thus if a is irrational, $|f(x) - f(a)| = |f(x)| < \frac{1}{n} < \varepsilon$ whenever x lies in the interval I_δ, i.e. whenever $|x - a| < \delta$. Hence f is continuous at a. When a is rational, $f(a) \neq 0$, but $\lim_{x \to 0} f(x) = 0$ by the same argument. Hence f is discontinuous at each rational point.

1. In each of the following cases explain whether the functions concerned can be made continuous at the point a, or identify the type of discontinuity. Sketch the graphs of the functions.

    (i) $x \mapsto \sqrt{x} \cos(\frac{1}{x})$, $\quad a = 0$;

    (ii) $x \mapsto \frac{x^3 - 8}{x^2 - 4}$, $\quad a = -2, a = 2$;

    (iii) $(x^2 - [x^2])$, $\quad a = n \in \mathbb{N}$ $\quad$ ($x \mapsto [x]$ is the integer part function).

2. Give a precise *definition* of the statement: 'the real function f is *discontinuous* at the point a in its domain'.

6.3 General power functions

We now embark on an extended example, which will illustrate, on the one hand, how a quite subtle concept of 'powers' can be built systematically from elementary ideas by using limits, and on the other, that our intuitive ideas of the properties of the power function can be justified in complete generality, provided always that we restrict attention to positive real numbers.

Before we can discuss the properties of the function $f_a : x \mapsto x^a$ for $x > 0$ and $a \in \mathbb{R}$, we need to decide what this function *means*: for example, what interpretation do we give to 3^π or $(\sqrt{2})^{\sqrt{2}}$? Clearly we cannot 'multiply 3 by itself π times', for example. In fact, there are even problems in defining rational powers: we know that negative numbers have no real square roots, but $(-2)^3 = -8$, so we can regard -2 as the *cube root* of -8, i.e. $(-8)^{1/3} = -2$. But since $\frac{1}{3} = \frac{2}{6}$, should we not also have $(-8)^{2/6} = -2$? This complicates matters when we try to interpret it as a *rational power*: on the one hand, we cannot find the sixth root of the negative number -8, so that we need to treat $(-8)^{2/6}$ as meaning $((-8)^2)^{1/6}$ (i.e. the sixth root of the square of -8). But on the other hand, $(-8)^2 = 64$, and this has sixth root equal to 2. (Nonetheless, defining odd roots of negative numbers does make sense, provided we always stick to writing the rational in lowest form – but we will not pursue this here.)

Rational powers

To avoid any difficulties we now confine ourselves to $x \geq 0$. In that case, the function $f_n : x \mapsto x^n$ is *one-one* for every $n \in \mathbb{N}$: to see this note that

$$y^n - x^n = (y - x)(y^{n-1} + y^{n-2}x + y^{n-3}x^2 + \ldots + x^{n-1})$$

so that $0 \leq x < y$ implies $0 \leq x^n < y^n$, i.e. f_n is *strictly increasing* on $[0, \infty)$, so that distinct points must map to distinct images. Thus the *inverse* function f_n^{-1} is well-defined on the *range* of f_n and we can *define* $x \mapsto x^{1/n}$ as this function f_n^{-1}. Now we can further define, for any positive rational $r = \frac{m}{n}$ (where, as usual, m and n are taken to have no common factors), f_r as the *composition* of f_m and f_n^{-1}, that is: given $x \geq 0$ compute $x^{1/n}$ and then find the mth power of this number, i.e.

$$x \mapsto x^{1/n} \mapsto (x^{1/n})^m = x^r = f_r(x)$$

(Since $x \geq 0$ and $r \geq 0$ this yields the same result as $(x^m)^{1/n}$.)

So $x \mapsto x^r$ makes sense for rational $r \geq 0$, and for $r < 0$ we can define x^{-r} first, and then make $x^r = \frac{1}{x^{-r}}$ for any $x > 0$. (Note that we must also omit the point 0 from the domain of f_r for negative $r \in \mathbb{Q}$.)

General powers

How should we define $f_a : x \mapsto x^a$ in general for the real number a? Again, we can first consider $a > 0$. It turns out to be advantageous first to define $g_a : x \mapsto a^x$ for such a and $x \in \mathbb{R}$. (This will provide a definition for such things as $2^{\sqrt{2}}$, for example, and at the same time introduce a whole class of functions which are continuous by their very definition.) Our construction proceeds in a number of distinct steps:

(i) To define $g_a(x) = a^x$ we first take $x = \frac{m}{n}$, so that $a^x = (a^{1/n})^m$ as just defined. This equals $(a^m)^{1/n}$, since $(a^x)^n = (a^{1/n})^{mn} = ((a^{1/n})^n)^m = a^m$ (using the power rule for *integer* powers, which is easily proved by induction), hence the (unique positive) n^{th} root of a^m is our a^x.

(ii) Next, we verify the power rule for *rational* powers: given $p = \frac{k}{l}, q = \frac{m}{n} \in \mathbb{Q}$, it is easy to see that

$$a^{p+q} = a^p a^q,$$

by considering the identities

$$(a^p a^q)^{ln} = (a^{k/l} a^{m/n})^{ln} = [(a^k)^{1/l}(a^m)^{1/n}]^{ln} = a^{kn} a^{ml} = a^{kn+ml}$$
$$= \left(a^{\frac{kn+ml}{ln}}\right)^{ln} = (a^{p+q})^{ln}$$

so that the result follows upon observing that the map $f_{ln} : x \mapsto x^{ln}$ is one-one for $x > 0$. Similarly, *you* should now show that $a^{p-q} = \frac{a^p}{a^q}$.

(iii) Now we temporarily restrict ourselves further by insisting that $a > 1$. The function $f_p : p \mapsto a^p$ is strictly increasing, since if $p < q$ are rationals, so that $q = p + r$ for some $r = \frac{m}{n} > 0$, then $a^q = a^{p+r} = a^p a^r = a^p (a^{1/n})^m > a^p$, since $a^{1/n} > 1$.

(iv) Finally, for $a > 1$, choose a sequence of rationals $r_n \to 0$ as $n \to \infty$. The crucial claim is that $a^{r_n} \to 1$.

To see this, let $\varepsilon > 0$ be given and find M so large that $M\varepsilon > a$. By the Bernoulli inequality we then have $(1 + \varepsilon)^M \geq M\varepsilon > a$. Consider a rational $r \in (-\frac{1}{M}, \frac{1}{M})$, so that $a^{-1/M} < a^r < a^{1/M}$, hence $a^r < a^{1/M} < 1 + \varepsilon$, while, on the other hand,

$$1 - \varepsilon = \frac{1 - \varepsilon^2}{1 + \varepsilon} < \frac{1}{1 + \varepsilon} < \frac{1}{a^{1/M}} = a^{-1/M} < a^r$$

Thus we have shown that $|a^r - 1| < \varepsilon$ for $|r| < \frac{1}{M}$. Now recall that our sequence $r_n \to 0$ as $n \to \infty$, hence we can find N large enough to ensure that $|r_n| < \frac{1}{M}$ for all $n > N$. Thus $|a^{r_n} - 1| < \varepsilon$ for $n > N$, so that $a^{r_n} \to 1$ as $n \to \infty$.

The hidden purpose behind these four steps is to allow us to make sense of the following *definition*: given $a > 1$ and $x \in \mathbb{R}$ we choose *any* sequence of rationals (x_n) which converges to x, and set $g_a(x) = a^x \equiv \lim_{n \to \infty} a^{x_n}$. Thus, for $a > 1$ at least, we finally have given a meaning to arbitrary *real* powers of a.

But this poses several questions:
- does the limit *exist?*

And if it does,
- is it *unique* – that is, is it the *same* whichever sequence in $\mathbb{Q}$, with limit x, we choose? Only then would we be able to claim that a^x is well-defined.
- And suppose that a^x is well-defined , can we extend 'basic properties', such as the power rule, to general a^x?
- Finally, what about the *function* $g_a : x \mapsto a^x$? Can we extend it to all $a > 0$? And is it one-one, so that we can find an *inverse* for it?

The next four steps answer these questions.

(v) First, suppose the sequence $(x_n) \subset \mathbb{Q}$ *increases* to $x \in \mathbb{R}$. Then $x = \sup_n x_n$ and we can find $k \in \mathbb{N}$ with $k > x$. Since $x_n \leq x_{n+1} < k$ we must have $a^{x_n} \leq a^{x_{n+1}} < a^k$ (since the x_n are rational, (iii) applies to them!) and hence the sequence $(a^{x_n})_n$ is an increasing sequence of real numbers which is bounded above by the real number a^k. Thus $\lim_{n \to \infty} a^{x_n}$ exists in $\mathbb{R}$, so that our definition of a^x *does* define a real number for such (x_n).

(vi) To see that the limit defining a^x is unique, let (x_n) be as in (v) and let (y_n) be *any* sequence of rationals converging to x. We need to show that (y_n) *also* defines a^x, i.e. that

$$\lim_{n \to \infty} a^{y_n} = \lim_{n \to \infty} a^{x_n}$$

To do this, we use (iv). Since (y_n) converges it must be bounded. Hence (a^{y_n}) is bounded, i.e. there is $K > 0$ such that $|a^{y_n}| < K$ for each $n \geq 1$. Now consider $r_n = x_n - y_n$, which is a *null* sequence of rationals, so that $a^{r_n} \to 1$ as $n \to \infty$, by what we proved in (iv). Thus we can write

$$a^{y_n} = a^{y_n} - a^{x_n} + a^{x_n} = a^{y_n}(1 - a^{r_n}) + a^{x_n}$$

Hence we have:

$$|a^{y_n} - a^x| \leq |a^{y_n}(1 - a^{r_n}) + a^{x_n} - a^x|$$
$$\leq |a^{y_n}||1 - a^{r_n}| + |a^{x_n} - a^x| < K|1 - a^{r_n}| + |a^{x_n} - a^x|$$

On the right we have $a^{r_n} \to 1$ and $a^{x_n} \to a^x$, so both sides tend to 0 as $n \to \infty$. Hence $\lim_{n \to \infty} a^{y_n} = a^x$.

This was the key step, since now $g_a : x \mapsto a^x$ is well-defined for $a > 1$ and *every* real x. It is easy to extend this to $a \in (0, 1]$: the case $a = 1$ is trivial: just set $1^x = 1$ for all $x \in \mathbb{R}$. If $0 < a < 1$ then $b = \frac{1}{a} > 1$, so that b^x is well-defined by the above steps. Now set $a^x = \frac{1}{b^x}$. Hence for all $a > 0$ the *power function* $g_a : x \mapsto a^x$ is defined for all $x \in \mathbb{R}$, that is, g_a has domain $\mathbb{R}$.

(vii) We verify that the extended function still satisfies the power rules:

$$a^x a^y = a^{x+y}, \quad \frac{a^x}{a^y} = a^{x-y}, \quad (a^x)^y = a^{xy}$$

Observe that if rationals $x_n \to x$ and $y_n \to y$ then $x_n + y_n \to x + y$, $x_n - y_n \to x - y$ and $x_n y_n \to xy$ as $n \to \infty$. Hence, by definition, we have $a^{x_n + y_n} \to a^{x+y}$, $a^{x_n - y_n} \to a^{x-y}$ and $a^{x_n y_n} \to a^{xy}$. On the other hand, by (ii), $a^{x_n + y_n} = a^{x_n} a^{y_n}$ and the product on the right has limit $a^x a^y$; and similarly, $a^{x_n - y_n} = \frac{a^{x_n}}{a^{y_n}}$, which has limit $\frac{a^x}{a^y}$, since the denominators are never 0. Finally,

$(a^{x_n})^{y_n} = a^{x_n y_n}$ for all n, hence $\lim_{n \to \infty} (a^{x_n})^{y_n} = a^{xy}$, while $a^{x_n} \to a^x$ implies that this limit is also $(a^x)^y$.

As limits are unique when they exist, we have verified the power rules for g_a.

(viii) To show that g_a is one-one, we first consider $a > 1$: if $x < y$ then $\varepsilon = \frac{1}{2}(y - x) > 0$ and we can choose a decreasing sequence (r_n) in $\mathbb{Q}$ converging to $2\varepsilon = y - x$, as well as an increasing sequence (q_n) in $\mathbb{Q}$ converging to ε. Then $q_n < r_n$ for all $n \geq 1$, so by (iii) $a^{q_n} < a^{r_n}$ for all n, hence also $a^{y-x} = a^{2\varepsilon} = \lim_{n \to \infty} a^{r_n} \geq \lim_{n \to \infty} a^{q_n} = a^\varepsilon > 1$. Thus we have shown $a^y > a^x$, so g_a is strictly increasing when $a > 1$. For $0 < a < 1$ we have now shown that g_b is strictly increasing, where $b = \frac{1}{a}$, and as $g_a(x) = \frac{1}{b^x}$ it follows that g_a is strictly decreasing when $0 < a < 1$. In *both* cases g_a is therefore one-one. The graphs are given below.

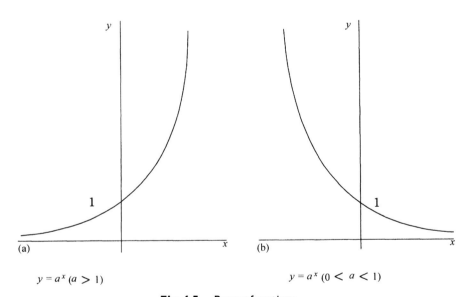

$y = a^x \ (a > 1)$ $y = a^x \ (0 < a < 1)$

Fig. 6.5. Power functions

The *inverses* can be sketched immediately by reflection of g_a in the line $y = x$.

Logarithms and exponentials

The function g_a^{-1} will be denoted by $\log_a$, and for $a > 1$ it is again strictly increasing; while for $0 < a < 1$ it is strictly decreasing. Its domain equals the *range* of g_a; while it seems intuitively obvious that this range is $(0, \infty)$, the proof of that claim has to be delayed for a while.

Our definitions show: $y = g_a(x) = a^x$ iff $x = g_a^{-1}(y) = \log_a(y)$. This allows us finally to deduce the familiar properties of the logarithm (note that this holds for *any base* $a > 0$). Suppose that $x, y > 0$.

(ix) $\log_a(xy) = \log_a x + \log_a y$; $\log_a(\frac{x}{y}) = \log_a x - \log_a y$.

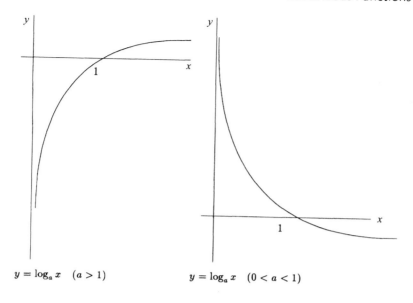

$y = \log_a x \quad (a > 1)$ $\qquad\qquad$ $y = \log_a x \quad (0 < a < 1)$

Fig. 6.6. Logarithmic functions

To see this, simply write $\alpha = \log_a x = g_a^{-1}(x)$ and $\beta = \log_a y = g_a^{-1}(y)$, so that $x = a^\alpha$, $y = a^\beta$, hence $xy = a^{\alpha+\beta}$ and $\frac{x}{y} = a^{\alpha-\beta}$ by the power rules in (vii). This just means that $\alpha + \beta = \log_a(xy)$ and $\alpha - \beta = \log_a(\frac{x}{y})$, as required.

Note that we have defined logarithms for all bases $a > 0$ in one fell swoop. In particular, this includes the bases 2, e, and 10, which you have seen before in other guises. Although our construction is rather long, each step follows fairly naturally from what came before. Now, however, we shall anticipate a later result: namely, that the range of g_a is indeed $(0, \infty)$ for each $a > 0$, so that $\log_a : (0, \infty) \mapsto \mathbb{R}$, as we expect to find from the graphs. In particular, take the base $e = \sum_{n=0}^\infty \frac{1}{n!}$ and consider the power function $g_e(x) = e^x$. Assuming that this function has range $(0, \infty)$ we can always find, for any given $a > 0$, a unique $b \in \mathbb{R}$ such that $a = e^b$. Thus $b = \log_e a$. (From now on $\log_e$ will just be written as $\log$.) Now consider $y = a^x$ for any real x: we have $y = (e^b)^x = e^{bx} = g_e(bx) = g_e(x \log a)$. Thus we have written $g_a(x) = a^x$ in the form $g_e(x \log a)$ for any real x and $a > 0$.

Finally, we turn all this round by restricting to *positive* x and exhanging the roles of x and a: this is how we will achieve our original objective, namely the extension to real powers of $f_n : x \mapsto x^n$ (!)

The map $f_a : x \mapsto x^a$ is therefore *defined* for positive x and *any* $a \in \mathbb{R}$, as $x \mapsto g_e(a \log x) = e^{a \log x}$. Our calculations above show that this function inherits the properties of $f_r : x \mapsto x^r$ for $r \in \mathbb{Q}$. Informally, we can now sketch its graph, and identify *four cases*: $a < 0$, $0 < a < 1$, $a = 1$, $a > 1$.

Since we defined e as $\sum_{n=0}^\infty \frac{1}{n!}$, it remains frustrating that we have not yet been able to link the power function g_e to the function exp whose values are given by the sum of the power series $\sum_{n=0}^\infty \frac{x^n}{n!}$. It is clear that these 'should' be the same function. To prove it we still need to establish the *continuity* of the function exp, to which we

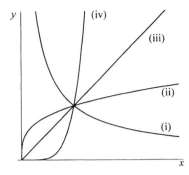

Fig. 6.7. The power function $y = x^a$ for different values of a: (i) $a < 0$, (ii) $0 < a < 1$, (iii) $a = 1$, (iv) $a > 1$

now turn our attention, and at the same time handle the continuity of power series in general.

EXERCISES ON 6.3

1. Suppose the real function f has domain $\mathbb{R}$ and $f(1) = a > 0$. Suppose further that $f(r).f(s) = f(r + s)$ for all rationals r, s. Show that $f(n) = a^n$ for all integers n, and then that $f(r) = a^r$ for all $r \in \mathbb{Q}$. Finally, show that if this function f is *continuous* then $f(x) = a^x$ for all real x.
2. Using the definition $x^a = e^{a \log x}$ show that $x^a . x^b = x^{a+b}$ and $(x^a)^b = x^{ab}$ whenever $x > 0$ and $a, b \in \mathbb{R}$.

6.4 Continuity of power series

To increase the stock of continuous functions at our disposal we now need to consider functions we have defined by power series, such as exp, sin, cos, cosh, etc. We know that these series define real functions for all $x \in \mathbb{R}$, since in each case we showed that the radius of convergence is infinite, so that the series converge absolutely for all x. But that does not *automatically* make them continuous: consider the geometric series defining $f(x) = \sum_{n=0}^{\infty} \frac{x^2}{(1+x^2)^n}$, i.e.

$$f(x) = x^2 + \frac{x^2}{1 + x^2} + \frac{x^2}{(1 + x^2)^2} + \ldots + \frac{x^2}{(1 + x^2)^n} + \ldots$$

For $x \neq 0$ the series has common ratio $\frac{1}{1+x^2} < 1$ and thus converges to $x^2(\frac{1}{1-\frac{1}{1+x^2}}) = 1 + x^2$, while it is obviously 0 at $x = 0$. Hence $f(0) = 0$, and $f(x) = 1 + x^2 > 1$ for all $x \neq 0$, so that f cannot be continuous at $x = 0$.

Now think of each *term* of the series as a function of x, i.e. $f_0(x) = x^2$, $f_1(x) = \frac{x^2}{1+x^2}, f_2(x) = \frac{x^2}{(1+x^2)^2}$, etc. Then Theorem 1 allows us to deduce that, for each fixed $k \in \mathbb{N}$,

$$\lim_{x \to a} \sum_{n=0}^{k} f_n(x) = \sum_{n=0}^{k} \lim_{x \to a} f_n(x)$$

so that the sum and limit operations are interchanged. However, the above example shows that we cannot replace k by ∞ without some restrictions on the functions (f_n). Fortunately, power series are better behaved.

• *Theorem 4* ————————

Suppose the power series $\sum_{n\geq 0} a_n x^n$ has radius of convergence $R > 0$. Then the function $f(x) = \sum_{n=0}^{\infty} a_n x^n$ defined by the series is continuous on the interval of convergence $(-R, R)$.

PROOF

Fix $x_0 \in (-R, R)$. Then we can find $0 < r < R$ such that $x_0 \in [-r, r]$, and for all $x \in [-r, r]$ we have $|a_n x^n| \leq |a_n| r^n$. Since $r < R$ we know that $\sum_{n\geq 0} |a_n| r^n$ converges. Thus, given $\varepsilon > 0$, we can find N such that the 'tail series' $\sum_{n>N} |a_n| r^n < \frac{\varepsilon}{3}$. Now for each $n \leq N$ we can use the continuity of $x \mapsto x^n$ to find $\delta_n > 0$ such that $|a_n||x^n - x_0^n| < \frac{\varepsilon}{3N}$ whenever $x \in N(x_0, \delta_n)$. Let $\delta = \min_{n\leq N} \delta_n$, then $\delta > 0$, and all these inequalities are valid simultaneously for $x \in N(x_0, \delta)$. Fix such an x. Write $s_N(x) = \sum_{n=0}^{N} a_n x^n$, and similarly for $s_N(x_0)$. Then $f(x) - f(x_0) = (s_N(x) - s_N(x_0)) + (f(x) - s_N(x)) + (f(x_0) - s_N(x_0))$ and so, by the Triangle Inequality:

$$|f(x) - f(x_0)| \leq |s_N(x) - s_N(x_0)| + |f(x) - s_N(x)| + |f(x_0) - s_N(x_0)|$$

But the first terms can be broken up further, into

$$\left| \sum_{n=0}^{N} a_n (x^n - x_0^n) \right| \leq \sum_{n=0}^{N} |a_n||x^n - x_0^n| < \sum_{n=0}^{N} \frac{\varepsilon}{3N} = \frac{\varepsilon}{3}$$

while, each of the last two terms is a 'tail series', and since both $|x|$ and $|x_0|$ are less than r, each of these two series is less than $\sum_{n>N} |a_n| r^n < \frac{\varepsilon}{3}$. Hence we have shown that $|f(x) - f(x_0)| < \varepsilon$ whenever $x \in N(x_0, \delta)$. Thus f is continuous at x_0, as required.

▣ *Example 8—Why* exp *really is 'exponential':* g_e = exp

Recall that $\exp(x) = \sum_{n=0}^{\infty} \frac{x^n}{n!}$ is defined for all $x \in \mathbb{R}$, and that $\exp(x + y) = \exp(x).\exp(y)$ for all $x, y \in \mathbb{R}$, as we saw in Chapter 4. Iterating, we have

$$\exp(x_1 + x_2 + \ldots + x_n) = \exp(x_1)\exp(x_2)\ldots\exp(x_n)$$

for any $x_1, x_2, \ldots, x_n \in \mathbb{R}$. In particular, letting each $x_i = 1$, we see that $\exp(n) = e^n = g_e(n)$, since $\exp(1) = e$. Hence the functions exp and g_e coincide for all $n \in \mathbb{N}$.

We extend this to positive rational powers, since g_e is one-one: let $r = \frac{m}{n}$ with $m, n \in \mathbb{N}$, then

$$[\exp(r)]^n = \exp(nr) = \exp(m) = g_e(m)$$

so that $g_e(r) = [g_e(m)]^{\frac{1}{n}} = \exp(r)$. As $\exp(0) = 1$ it follows from this and the identity $\exp(x).\exp(-x) = \exp(0)$ that $\exp(-r) = e^{-r}$, so that exp and g_e coincide on $\mathbb{Q}$.

Moreover, exp is clearly strictly increasing on $(0, \infty)$, i.e. if $0 < x < y$ then $0 < \exp(x) < \exp(y)$. We can now see that $\exp(-y) < \exp(-x)$ also, and that exp is strictly positive and increasing on $\mathbb{R}$. But it is also continuous on $\mathbb{R}$, by Theorem 4, and coincides with the continuous function g_e on $\mathbb{Q}$. By Exercises on 6.1(2) it follows that these two functions coincide on $\mathbb{R}$, justifying our use of $g_e(x) = e^x$ for $\exp(x)$ in all cases.

It is immediately obvious that Theorem 4 also guarantees the continuity of the trigonometric and hyperbolic functions defined via power series in Chapter 4. The link of sin and cos, as defined by power series, with ratios of sides of right-angled triangles is not quite so quickly established – some of the details are left for a Tutorial Problem in the next chapter.

⊛ *Example 9—Abel's Theorem*

If the power series $\sum_{n \geq 0} a_n x^n$ has radius of convergence 1 and the series $\sum_{n \geq 0} a_n$ converges (that is, the power series also converges at $x = 1$) then the function f defined by the series is continuous at $x = 1$. In other words:

$$\lim_{x \to 1} f(x) = \sum_{n=0}^{\infty} a_n$$

This theorem was first proved by Abel in the 1820s. Using Theorem 4 we can provide a concise proof:

Write $s_n = a_0 + a_1 + \ldots + a_n$ for the partial sums of the series $\sum_n a_n$ and set $s_{-1} = 0$ for convenience. Then $a_n = s_n - s_{n-1}$ for all n and we have:

$$\sum_{n=0}^{k} a_n x^n = \sum_{n=0}^{k} (s_n - s_{n-1}) x^n = (1 - x) \sum_{n=0}^{k} s_n x^n + s_k x^k$$

When $x \in (-1, 1)$ we let $k \to \infty$, so that the left-hand side tends to $f(x)$, while on the right $(s_k x^k)$ is null, since (s_k) converges to a finite limit s (and hence is bounded) and (x^k) is null. Thus

$$f(x) = (1 - x) \sum_{n=0}^{\infty} s_n x^n$$

Given $\varepsilon > 0$ choose N so that $|s - s_n| < \frac{\varepsilon}{2}$ for $n > N$. Then, because $(1 - x) \sum_{n=0}^{\infty} x^n = 1$ when $|x| < 1$, we can write

$$|f(x) - s| = |(1 - x) \sum_{n=0}^{\infty} (s_n - s) x^n| \leq |1 - x| \sum_{n=0}^{N} |s_n - s| |x|^n + \frac{\varepsilon}{2}$$

and this can be made less than ε by choosing x close enough to 1. This completes the proof.

Abel's Theorem can be used, for example, to prove that whenever the Cauchy product of two series converges, then its sum is the product of the sums of the two series – here no assumption about absolute convergence is required. See Chapter 8 of W. Rudin's classic text *Principles of Mathematical Analysis* for more details.

EXERCISE ON 6.4

1. Using the series definitions of exp and sin and the continuity of power series show that

$$\lim_{x \to 0} \frac{\exp x - 1}{x} = 1 = \lim_{x \to 0} \frac{\sin x}{x}$$

Summary

The continuity of a real function at a point a in its domain was discussed as a *local* property of the function; that is, the definition requires information about the function only for points near a. This can be expressed simply in terms of limits of functions (and hence of sequences) – which again allows us to deduce stability properties of the set of all continuous functions in a simple manner – but also in terms of the concept of a *neighbourhood* of the point a.

The link with limits provides a simple classification of reasons why a function can fail to be continuous at a given point – this shows how the connection between the formal definition of continuity and its graphical representation becomes rather tenuous when we consider functions with complicated domains.

Since most of our examples will in fact be functions defined on intervals – to be analysed more precisely in the next chapter – the last two sections concentrate on two such cases: power functions with general exponents, and the continuous functions defined by power series within their interval of convergence.

FURTHER EXERCISES

1. Suppose that the real function f has the following property: there exists $M > 0$ such that for all $x, x' \in \mathbb{R}$,

   $$|f(x) - f(x')| \le M|x - x'|$$

   Show that f is a continuous function.
2. Suppose the real function f is continuous at $a \in \mathbb{R}$ and $f(x) < 0$ for all $x < a$, $x \in \mathcal{D}_f$, while $f(x) > 0$ for all $x > a$, $x \in \mathcal{D}_f$. Show that $f(a) = 0$.
3. Explain carefully why the functions sinh and cosh are continuous on $\mathbb{R}$.
4. By comparing the power series $\sum_{n \ge 0} x^n$ and $\sum_{n \ge 0}^{\infty} \frac{x^n}{n!}$ show that $1 + x \le \exp x \le (1 - x)^{-1}$ whenever $|x| < 1$.

7 • Continuity on Intervals

In the previous chapter we considered several examples to show that the 'obvious' definition of continuity is not sufficiently precise to deal with all the situations we wish to handle. In particular, we concentrated on continuity as a *local* property of a function; that is, the continuity or otherwise of f at the point a is determined by the behaviour of the function *in a neighbourhood of a*. Thus, for example, we are able to conclude that $f(x) = \frac{1}{x}$ is in fact continuous throughout its domain, since it is undefined at 0 and any point $a \neq 0$ has a neighbourhood which *excludes* 0.

This example contradicts our naive idea that the graph of a continuous function 'is in one piece', since the domain of $f(x) = \frac{1}{x}$ is not even in one piece! However, if we restrict ourselves to functions whose **domain** consists of one piece, i.e. is an *interval*, then we can again ask how accurate our intuitive idea of continuity was. And this time it will turn out to be much closer to the truth, so that our basic theorems will at first sight appear to be 'stating the obvious' and so be hardly worth proving – when we apply them, however, their importance and usefulness will become clear.

7.1 From interval to interval

Recall that the fundamental property that makes a set I an *interval* is that, given any two points $a, b \in I$, all points between a and b also belong to I. So: if $a \leq c \leq b$ then $c \in I$. Suppose $f : I \mapsto I$ is given and $[a, b] \subset I$, with $f(a) \neq f(b)$. Does f take on all values *between* a and b? In other words, will the *image* under f of an interval again be an interval? Our first theorem says that it will, provided that f is continuous on I. (To be a little more precise about the behaviour of f at the *endpoints* of the interval if $I = [a, b]$: we shall assume that f is continuous on the right at the left endpoint of I and continuous on the left at the right endpoint of I.)

⊛ Example I

That our claim is not quite as obvious as it seems becomes clearer if we consider $f(x) = x^2$ on the set $\mathbb{Q}$ rather than $\mathbb{R}$: this function is certainly continuous at each point of $\mathbb{Q}$, and on $\mathbb{Q} \cap [0, 1]$ it takes on all values of the form r^2 for $r \in \mathbb{Q} \cap [0, 1]$, including 0 and 1. But it never takes the value $\frac{1}{2} \in \mathbb{Q} \cap [0, 1]$, since $\frac{1}{\sqrt{2}}$ is not rational. Hence our results depend directly on the *completeness of* $\mathbb{R}$, and cannot be transferred to functions defined only at rational points.

● Theorem I—Intermediate Value Theorem ————————

Suppose that $f : I \mapsto I$ is continuous and $a < b$ belong to I, with $f(a) < f(b)$. For each value $\gamma \in (f(a), f(b))$ there exists $c \in (a, b)$ for which $f(c) = \gamma$.

PROOF

Since $f(a) < \gamma < f(b)$, the set $A = \{x \in [a, b] : f(x) < \gamma\}$ is non-empty and bounded above, and therefore has a supremum $c \in \mathbb{R}$. We shall show that $f(c) = \gamma$; in the process we shall find a use for both versions of the definition of continuity.

Given any $\varepsilon > 0$ we can find $x \in A$ with $x + \varepsilon > c$, since c is the *least* upper bound of A. On the other hand, for sufficiently small $\varepsilon > 0$ the point $y = c + \varepsilon$ must belong to $[a, b] \setminus A$: to see this, recall that $\gamma < f(b)$, so that if $\varepsilon < f(b) - \gamma$, the left-continuity of f at b provides $\delta > 0$ such that $f(x) \in N((f(b), \varepsilon)$ when $b - \delta < x < b$. But then $f(x) > \gamma$ for all such x (e.g. $x = b - \frac{\delta}{2}$), and so $c < b$.

Apply the above successively with $\varepsilon = \frac{1}{n}$ (starting with n large enough to ensure that $y_n \in [a, b]$ for all n); we define sequences (x_n) and (y_n) of points of A and $[a, b] \setminus A$ respectively, both converging to c. Since f is continuous at $c \in [a, b]$, it follows that $f(x_n) \to \gamma$ and $f(y_n) \to \gamma$ as $n \to \infty$. But since $x_n \in A$ we have $f(x_n) < \gamma$, while $f(y_n) \geq \gamma$, as $y_n \notin A$. Now apply Proposition 4 of Chapter 2: the sequence (x_n) ensures that $f(c) \leq \gamma$ and (y_n) that $f(c) \geq \gamma$; hence $f(c) = \gamma$.

● Corollary

If I is an interval and $f : I \mapsto I$ is continuous, then $f(I)$ is also an interval.

PROOF

Note that this contains the trivial case of a constant function f, where for any two points $a, b \in I$ we have $f(a) = f(b) = \gamma$ (say). In this case $f(I) = \{\gamma\}$, which is the 'interval' $[\gamma, \gamma]$. For non-constant f take any two points of I with distinct images $\alpha < \beta$. Then Theorem 1 applies, and we can conclude that any $\gamma \in (\alpha, \beta)$ is the image of some $c \in I$, hence belongs to $f(I)$, which is therefore an interval.

TUTORIAL PROBLEM 7.1

The proof of the Intermediate Value Theorem (henceforth abbreviated to *IVT*) contains an application of completeness which deserves further comment: the reason why we can extract sequences (x_n) in A and (y_n) in $B = [a, b] \setminus A$, both converging to c, is that these sets 'meet' at c. This is a special case of the following general principle, which you should explore:

Given two disjoint sets A, B whose union is an interval, then B contains a point *at zero distance* from A; that is, $d(b, A) := \inf\{|b - x| : x \in A\} = 0$.

How can this be used to prove the *IVT*?

Our next result will show, in particular, that $\log_a$ is also continuous on its domain.

● Theorem 2

If f is continuous and strictly increasing [resp. decreasing] on $[a, b]$ then the inverse function f^{-1} exists and is strictly increasing [resp. decreasing] and continuous on the closed interval with endpoints $f(a)$ and $f(b)$.

PROOF

Since f is strictly monotone on $[a, b]$ it is one-one on this interval, and so f^{-1} exists. By the *IVT* its domain is $I = f([a, b]) = [f(a), f(b)]$, if f is continuous and strictly increasing on $[a, b]$, and $[f(b), f(a)]$ in the other case. It is easy to check directly that f^{-1} is strictly increasing [resp. decreasing] if f is.

To show that f^{-1} is continuous we can apply the Bolzano–Weierstrass Theorem of Chapter 3. Let y be a point in the interval I, and suppose that $y = f(x)$. Also let $y_n = f(x_n)$ define a sequence $y_n \to y$ as $n \to \infty$. We need to show that $x_n \to x$ as $n \to \infty$. First we must prove that (x_n) converges at all. Suppose, therefore, that (x_n) diverges. Since $(x_n) \subset [a, b]$ is bounded, Proposition 3 in Chapter 3 shows that there are at least two subsequences with different limits, say $x_{m_r} \to L$ and $x_{n_r} \to L'$ as $r \to \infty$. Since f is continuous, $y_{m_r} = f(x_{m_r}) \to f(L)$ and $y_{n_r} = f(x_{n_r}) \to f(L')$. But $y_n \to y$, hence all subsequences converge to y also. Therefore $f(L) = y = f(L')$, which contradicts the fact that f is one-one on $[a, b]$. Hence (x_n) converges.

Now if $\lim_{n \to \infty} x_n = z$, then $y_n = f(x_n) \to f(z)$, so that $y = f(z)$ by the uniqueness of limits. Again because f is one-one it follows that $z = x$, which completes the proof.

● *Example 2*

Theorem 1 and Theorem 2 allow us to complete our discussion of the power functions $g_a : x \mapsto a^x$ (defined for $a > 0$ and with domain $\mathbb{R}$) and $f_a : x \mapsto x^a$ (with domain $[0, \infty)$ when $a \geq 0$, and $(0, \infty)$ when $a < 0$): the continuity of g_a was proved in Chapter 6, and now Theorem 2 has shown that its inverse $\log_a$ is also continuous; all that is still needed to justify the graphs given in Chapter 6 is to identify the *range* of each function.

First consider $g_a : x \mapsto a^x$: the function is continuous on the *interval* $\mathbb{R}$ by construction (as we saw in Chapter 6); this is now also evident from the representation $g_a(x) = \exp(x \log a)$ and Theorem 4 in Chapter 6 (recall that $\log = \log_e$ by our convention!). Hence by the *IVT* we conclude that $\mathcal{R}_{g_a}$ is an interval. Since $a > 0$, 0 is a lower bound for the range. To show that in fact $\mathcal{R}_{g_a} = (0, \infty)$, first assume that $a > 1$. Then $a = (1 + h)$ for some positive h, and $a^n = (1 + h)^n \geq 1 + nh$ by the Bernoulli inequality, for all $n \in \mathbb{N}$. Given any $K > 0$ we can always find $n \in \mathbb{N}$ such that $n > \frac{K}{h}$, hence $a^n > K$. This shows that the range of g_a is unbounded above. Similarly, the range of g_a contains $a^{-n} = (1 + h)^{-n} \leq \frac{1}{nh}$, so that 0 is in fact the greatest lower bound of $\mathcal{R}_{g_a}$. Hence if $a > 1$ we have shown that $\mathcal{R}_{g_a} = (0, \infty)$, as claimed.

In particular, the exponential function $g_e = \exp$ has range $(0, \infty)$. Note that in this case we can write down the limiting values at once:

$$\lim_{x \to \infty} \exp(x) = \infty, \quad \text{since } \exp(x) = \sum_{n=0}^{\infty} \frac{x^n}{n!} > x \text{ for positive } x$$

while

$$\lim_{x \to -\infty} \exp(x) = 0, \quad \text{since } \exp(-x) = \frac{1}{\exp(x)}$$

For $0 < a < 1$ the function $g_a(x) = a^x = (\frac{1}{b})^x = \frac{1}{b^x}$ for $b = \frac{1}{a} > 1$. Since we have just shown that g_b has range $(0, \infty)$, so does g_a.

In each of the above cases the inverse function $g_a^{-1} = \log_a$ thus has domain $(0, \infty)$ and range $\mathbb{R}$.

Finally, consider $x \mapsto x^a$ for $x > 0$: in that case we can use the representation $\exp(a \log x)$ to confirm that the function is continuous on the interval $(0, \infty)$. Hence by the *IVT* its range is an interval, and since $\log x$ takes on all real values as x ranges through $(0, \infty)$ it follows that this interval is again $(0, \infty)$. Now for $a < 0$ the (strictly *decreasing*!) function $f_a : x \mapsto x^a$ is defined only for $x > 0$, hence $\mathcal{R}_{f_a} = (0, \infty)$. For $a = 0$ we obtain the constant function $f_0(x) = 1$, and for $a > 0$ f_a has domain $[0, \infty)$, since $f_a(0) = 0^a = 0$ is well-defined. In this case f_a is strictly *increasing* and $\mathcal{R}_{f_a} = [0, \infty)$. (Study the graphs given in Chapter 6 once more!)

⊕ *Example 3*

The value of the representations $x \mapsto \exp(x \log a)$ and $x \mapsto \exp(a \log x)$ will become evident when we use the operations of the Calculus on the functions g_a and f_a. We can see evidence of this already by considering the function $h : x \mapsto x^x$ for $x > 0$. Since $h(x) = \exp(x \log x)$ on the domain $(0, \infty)$ of this continuous function, its range is again an interval. This interval is unbounded above, since $x \log x \to \infty$ as $x \to \infty$. On the other hand, $h(1) = \exp(0) = 1$, and $h(x) \to 1$ as $x \downarrow 0$, since $x \log x \to 0$ as $x \downarrow 0$ (to see this, write $x = e^t$ and let $t \to -\infty$, then note that $te^t \to 0$ as $t \to -\infty$, by Exercises on 4.4). Hence h must have a turning point (minimum) in the interval $(0, 1)$. To find this minimum point we shall need to calculate the derivative of h. This is deferred to the next chapter.

TUTORIAL PROBLEM 7.2

We have not yet defined π without reference to geometry. The following steps explore how one may give 'analytic' definition of π – you should supply the details.

First derive the addition formula

$$\cos(x + y) = \cos x \cos y - \sin x \sin y$$

from the series definitions of cos and sin, using arguments similar to those of Tutorial Problem 4.4.

Deduce that $\cos x - \cos y = 2 \sin(\frac{x+y}{2}) \sin(\frac{y-x}{2})$.

Next, use the first few terms of the series defining sin to show that $\sin x > 0$ on $(0, 2)$, and deduce from the above identity that cos is strictly decreasing on $[0, 2]$. Estimating the first few terms of the series defining cos, show that $\cos 2 < 0$. Now use the *IVT* to deduce that there is *exactly one* $x \in (0, 2)$ at which cos takes the value 0. *Define* this point as $\frac{\pi}{2}$, i.e. set $\pi = 2x$.

Deduce that $\sin(\frac{\pi}{2}) = 1$. The familiar properties of sin and cos can be deduced from these beginnings.

⦿ *Example 4*

Restricting the continuous function sin to the interval $[-\frac{\pi}{2}, \frac{\pi}{2}]$ we can define its inverse, since sin is strictly increasing on this interval. The inverse function $\sin^{-1} : [-1, 1] \mapsto [-\frac{\pi}{2}, \frac{\pi}{2}]$ is continuous by Theorem 1. Similar remarks apply to $\cos^{-1}$ and the other inverse trigonometric functions.

Our theorems also confirm why it appears to be very difficult to draw the graph of a continuous one-one function which is not strictly monotone (unless we allow the domain to have a 'gap', as with $f(x) = \frac{1}{x}$). When the domain is to be an interval, no such continuous function is possible!

● *Proposition 1*

If $f : [a, b] \mapsto \mathbb{R}$ is continuous and one-one, then f is strictly monotone on $[a, b]$, and f^{-1} is continuous on $f([a, b])$.

PROOF
If f were neither strictly increasing nor strictly decreasing on $[a, b]$, then we would be able to find $a_1 < a_2 < a_3$ in $[a, b]$ such that $f(a_1) < f(a_3) < f(a_2)$. Applying the *IVT* to f on $[a_1, a_2]$ we could then find $x \in (a_1, a_2)$ such that $f(x) = f(a_3)$. But $x < a_2 < a_3$, hence f would not be one-one. This proves that f is strictly monotone on $[a, b]$. The continuity of f^{-1} now follows from Theorem 2.

EXERCISES ON 7.1

1. Use the *IVT* to find the *image set $f(I)$* in each of the following cases (sketch the graphs!):
   (i) $f(x) = x + 2|x|$, $\quad I = [-2, 2]$
   (ii) $f(x) = \sin x - \cos x$, $\quad I = [0, \frac{\pi}{2}]$
2. Give an example of a strictly monotone function f defined on the open interval $(0, 1)$ which is not continuous on $(0, 1)$. Is f^{-1} continuous on its domain for your example? Try to draw a general conclusion from your observations.

7.2 Applications: fixed points, roots and iteration

Having worked hard to establish our theorems we can now collect some of the results:

● *Proposition 2*

Let $f : [a, b] \mapsto [a, b]$ be a continuous function. Then f has a *fixed point* in $[a, b]$; that is, there is at least one $c \in [a, b]$ such that $f(c) = c$.

PROOF
We apply the *IVT* to the continuous function g defined by $g(x) = f(x) - x$. Since the range of f is contained in $[a, b]$, we must have $f(a) \geq a$ and $f(b) \leq b$. Hence $g(a) = f(a) - a \geq 0$, while $g(b) = f(b) - b \leq 0$. By the *IVT* there exists $c \in [a, b]$ such that $g(c) = 0$, i.e. $f(c) = c$.

This simple consequence of the *IVT* turns out to be quite difficult to generalize to higher dimensions – though it remains true in much greater generality.

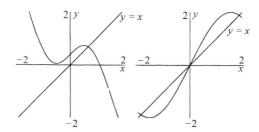

Fig. 7.1 Examples of fixed points: (a) unique fixed point, (b) multiple fixed points

⊛ *Example 5*

The fixed point need not be unique: let $f(x) = 2\sin(x)$, which maps the interval $[-2, 2]$ into itself. The graph above shows that f has *three* fixed points in this interval.

The proof of Proposition 2 indicates that the search for fixed points is just a special case of looking for 'roots' of functions, i.e. finding points at which a given continuous function takes the value 0. For polynomials we have a quite general result:

● *Proposition 3*

Every polynomial of odd degree has at least one real root.

PROOF
Suppose that $P(x) = a_0 + a_1 x + a_2 x^2 + \ldots + a_n x^n$, where $a_n \neq 0$. If $a_0 = 0$ then $x = 0$ is a root (i.e. $P(0) = 0$). Otherwise, for $x \neq 0$, divide through by a_n, to obtain $\frac{P(x)}{a_n} = x^n(1 + a_{n-1}x^{-1} + a_{n-2}x^{-2} + \ldots + a_0 x^{-n})$, which has the same roots as $P(x)$. Since n is odd, x^n is negative for $x < 0$ and positive for $x > 0$. For large $|x|$ the factor in the brackets tends to 1, so that if $a_n > 0$, $\frac{P(x)}{a_n}$ has the same sign as x, i.e. positive for large positive x, negative for large negative x. If $a_n < 0$ the situation is reversed. Hence in all cases $\frac{P(x)}{a_n}$, and thus $P(x)$, must take the value 0 at least once by the *IVT*.

⊛ *Example 6*

The polynomial $P(x) = x^5 - 3x + 1$ has odd degree, thus has a real root by Proposition 3. To locate such a root, we can again use the *IVT* by considering the sign of $P(x)$ for various values of x. For example, $P(0) = 1$ and $P(1) = -1$. Hence by the *IVT* $P(x) = 0$ for some $x \in (0, 1)$. We can repeat the process to decide which of the intervals $[0, \frac{1}{2}]$ and $[\frac{1}{2}, 1]$ contains a root: simply calculate $P(\frac{1}{2}) = \frac{1}{32} - 3(\frac{1}{2}) + 1$, which is negative, so that the *IVT* ensures that the interval $[0, \frac{1}{2}]$ contains a root. Similarly, we calculate $P(\frac{1}{4}) = \frac{1}{1024} - 3(\frac{1}{4}) + 1 > 0$, so that $(\frac{1}{4}, \frac{1}{2})$ contains a root, etc.

Similar arguments can be applied to more general functions: approximate solutions to the equation $3^x = 4x$ can be found as approximate roots of the continuous function $f(x) = 3^x - 4x$. Since $f(0) = 3^0 - 4(0) = 1$, $f(1) = 3^1 - 4(1) = -1$, the *IVT* guarantees a root $x \in (0, 1)$; and since $f(\frac{1}{2}) = \sqrt{3} - 2 < 0$, we can restrict attention further to $(0, \frac{1}{2})$. Of course, the calculations become more and more tedious as we narrow down the location of x, and we shall want to find a more effective *algorithm* for dealing with such situations.

We made some initial progress towards such an algorithm in Chapter 2, where we considered *recursive* sequences, i.e. sequences whose terms were produced by the *iterative* application of some function. We have noted already that if $x_n \to x$ and $x_{n+1} = f(x_n)$, then $f(x) = x$ *provided* that f is continuous. However, we still need to find conditions on f which guarantee that the iterative sequence (x_n) with $x_{n+1} = f(x_n)$ actually converges at all. This can only happen if the equation $f(x) = x$ *has* a solution, i.e. if $f(x) - x = 0$ for some real x. Thus, for example, since $x^2 + 1 = 0$ has *no* real roots, it follows that $f(x) = x^2 + x + 1$ is a continuous function for which the equation $f(x) = x$ has no solution. In particular, therefore, f cannot map any closed bounded interval $[a, b]$ into itself, by Proposition 2.

Even if for some continuous f the equation $f(x) = x$ has a solution $c \in \mathbb{R}$, and $x_{n+1} = f(x_n)$ begins at some x_0 near c, it does not necessarily follow that $x_n \to c$ as $n \to \infty$, since we cannot guarantee that (x_n) is even *defined* for all n; the iteration could take us outside $\mathcal{D}_f$ at some stage, and then could not be continued. Consider $f(x) = \frac{1}{\sqrt{1-x^2}} - 2$ and $x_0 = 0.95$, for example, where $x_1 > 1$, but $\mathcal{D}_f = (-1, 1)$.

Thus to be 'safe' we shall require that f maps some interval I to itself and $x_0 \in I$. This ensures that the equation $f(x) = x$ has a solution $c \in I$ (by Proposition 2), and the sequence (x_n) is now well-defined by taking $x_0 \in I$ and setting $x_{n+1} = f(x_n)$ for all $n \geq 0$. But even now we cannot be sure that $x_n \to c$. Take, for example, $x_0 = 1$ and $f(x) = -x$ on $I = [-1, 1]$, then $x_n = (-1)^n$, which diverges.

A natural idea for excluding such examples is to demand that applying f should bring points 'closer together': we call f a *contraction mapping* if it satisfies the following condition on the interval I: there exists $\alpha \in (0, 1)$ such that $|f(x) - f(y)| \leq \alpha|x - y|$ for all $x, y \in I$.

With this additional condition we can finally show that the iteration converges to a *unique* fixed point of f in I.

● *Proposition 4*

If $f : I \mapsto I$ is a contraction mapping on the closed bounded interval I and $x_0 \in I$, then the sequence (x_n) defined by $x_{n+1} = f(x_n)$ converges to the unique fixed point of f in I.

PROOF
Since any contraction mapping is continuous (for given $\varepsilon > 0$ take $\delta = \frac{\varepsilon}{\alpha}$ to see this), Proposition 6.2 guarantees the existence of at least one fixed point $c \in I$. This fixed point is unique, since if $c' \neq c$ also satisfies $f(c') = c'$, then

$$|c' - c| = |f(c') - f(c)| \leq \alpha|c' - c| < |c' - c|$$

which is impossible.

To see that the iterative sequence (x_n) converges, we apply the contraction property repeatedly: for each $n \geq 1$

$$|x_{n+1} - x_n| = |f(x_n) - f(x_{n-1})| \leq \alpha |x_n - x_{n-1}|$$

hence, generally,

$$|x_{n+1} - x_n| \leq \alpha^n |x_2 - x_1| \leq \alpha^n (b - a)$$

if we take $I = [a, b]$ (as we may). Thus, writing

$$s_n = x_n - x_0 = (x_n - x_{n-1}) + (x_{n-1} - x_{n-2}) + \ldots + (x_1 - x_0)$$

we can take s_n as the nth partial sum of the series $\sum_{n \geq 1}(x_n - x_{n-1})$, which is absolutely convergent by comparison with $\sum_{n \geq 1} \alpha^n$ (this is where we use the fact that $|\alpha| < 1$). This shows that $(x_n - x_0)_{n \geq 1}$ converges, and hence that (x_n) converges. Since each $x_n \in I$ and $I = [a, b]$ is a closed interval, it follows that $x = \lim_{n \to \infty} x_n$ is also in I. But $f(x) = x$ follows as before, so that x is a fixed point of f, and therefore $x = c$.

Example 7

The function $f(x) = \sqrt{x + 1}$ is a contraction map on $[0, 2]$, since for $0 \leq y < x$ we obtain

$$|f(x) - f(y)| = \sqrt{x + 1} - \sqrt{y + 1} = \frac{(\sqrt{x+1} - \sqrt{y+1})(\sqrt{x+1} + \sqrt{y+1})}{\sqrt{x+1} + \sqrt{y+1}}$$

$$= \frac{|x - y|}{\sqrt{x+1} + \sqrt{y+1}} < \frac{1}{2}|x - y|$$

so that we can take $\alpha = \frac{1}{2}$ in this case. This means that the sequence (x_n) with $x_1 \in [0, 2]$ and $x_{n+1} = f(x_n)$ converges to the unique fixed point x of f in $[0, 2]$, i.e. the solution of the equation $x = \sqrt{x + 1}$, which reduces to the quadratic $x^2 - x - 1 = 0$, whose solution in $[0, 2]$ is $x = \frac{1+\sqrt{5}}{2}$. (The 'golden section' – once again!)

TUTORIAL PROBLEM 7.3

The uniqueness of the fixed point requires the constant α to be less than 1, but, as we saw in Further Exercise 1 at the end of Chapter 6, any function which satisfies the condition $|f(x) - f(y)| \leq M|x - y|$ for some $M > 0$ on an interval I must be continuous on that interval. We call such conditions *Lipschitz conditions* and this rather special kind of continuity is known as Lipschitz continuity. Find out what you can about the use of these ideas in the iterative solution of differential equations.

EXERCISES ON 7.2

1. Suppose that $P_n(x) = a_0 + a_1 x + a_2 x^2 + \ldots + a_n x^n$ is a polynomial with *even* degree, and such that $a_0 < 0$, $a_n = 1$. Show that P_n has at least two real roots.

2. Use the identity $\cos x - \cos y = 2\sin(\frac{x+y}{2})\sin(\frac{y-x}{2})$ to show that cos is a contraction map on [0,1], and hence write down an iterative procedure for finding the unique solution in [0,1] of the equation $x = \cos x$. (To find $\alpha < 1$ you may need to 'cheat' slightly by looking up values for sin 1 and sin$\frac{1}{2}$; to remain honest you can use the fact that sin is defined by an alternating sequence, so that, for example, $\sin 1 < 1 - \frac{1}{6} + \frac{1}{120}$.)

3. Let the function $f(x) = (\cos x)^4$ be defined on [0,1]. Show that there exists a point $a \in [0, 1]$ such that $\cos a = \sqrt[4]{a}$.

7.3 Reaching the maximum: the Boundedness Theorem

Continuous functions on closed, bounded intervals, that is, of the form $[a, b]$, are especially 'well-behaved'. We have already exploited this in our discussion of iterative solutions of equations. The situation can be summarized quite simply:

● *Theorem 3—The Boundedness Theorem* ————————

Suppose that $f : [a, b] \mapsto \mathbb{R}$ is continuous. Then f is bounded and attains its bounds, so that the range $\mathcal{R}_f = [\alpha, \beta]$ is again a closed bounded interval.

(We say that f *attains its bounds* if there exist $c, d \in [a, b]$ such that $\alpha = \inf \mathcal{R}_f = \min \mathcal{R}_f = f(c)$ and $\beta = \sup \mathcal{R}_f = \max \mathcal{R}_f = f(d)$.)

PROOF

First we need to show that f is a bounded function. So suppose that $\mathcal{R}_f$ were unbounded above. Then we can choose a sequence of points $x_n \in [a, b]$ such that $f(x_n) > n$ for each $n \in \mathbb{N}$. But $(x_n) \subset [a, b]$ is a bounded sequence, hence by the Bolzano–Weierstrass Theorem we can find a convergent subsequence $(x_{n_r})_{r \geq 1}$ whose limit x is in the interval $[a, b]$. Since f is continuous at x, $f(x_{n_r}) \to f(x)$ as $r \to \infty$. On the other hand by construction of the sequence (x_n), $f(x_{n_r}) > n_r$ for each $r \geq 1$. This is impossible, since $n_r > |f(x)| + 1$ for large enough r. The contradiction shows that f is bounded above. A similar argument shows that f is bounded below.

To see that $\beta = \sup\{f(x) : x \in [a, b]\} = f(c)$ for some $c \in [a, b]$, we again argue by contradiction. If $\beta \neq f(x)$ for all $x \in [a, b]$, then $\beta > f(x)$ and hence $g(x) = \frac{1}{\beta - f(x)}$ is a well-defined continuous function on $[a, b]$. Hence by what has just been proved, g is also bounded on $[a, b]$, i.e. for some $K > 0$ we have $g(x) < K$ for all $x \in [a, b]$. But by the definition of sup, the denominator $\beta - f(x)$ can be made arbitrarily small, that is, we can find $x \in [a, b]$ such that $\beta - f(x) < \frac{1}{K}$. Hence $g(x) > K$, which is a contradiction. This shows that we cannot have $\beta > f(x)$ for all $x \in [a, b]$, and thus there must exist $c \in [a, b]$ such that $\beta = f(c)$. The existence of d such that $f(d) = \inf\{f(x) : x \in [a, b]\}$ is shown similarly.

Application: impossible continuous functions!

The two main theorems of this chapter (the Intermediate Value and Boundedness Theorems) effectively restrict the behaviour of continuous functions on intervals, and on closed bounded intervals in particular. This precludes 'unpleasant' func-

certainly take $y \in (0, \delta)$ such that $|f(x) - f(y)| = |\frac{1}{x} - \frac{1}{y}| > \varepsilon$, even though $|x - y| < \delta$. On the other hand, when the set A is a closed, bounded interval, then the two concepts coincide:

● *Theorem 4*

If $f : [a, b] \mapsto \mathbb{R}$ is continuous, then it is uniformly continuous on $[a, b]$.

PROOF

If f is *not* uniformly continuous on $[a, b]$ then there exists $\varepsilon > 0$ for which no 'corresponding $\delta > 0$' can be found; that is, given $\delta > 0$, we can always find a pair of points $x, y \in [a, b]$, depending on δ, such that $|x - y| < \delta$, but $|f(x) - f(y)| \geq \varepsilon$. Use this successively with $\delta = 1, \frac{1}{2}, \ldots, \frac{1}{n}, \ldots$, to define a pair of sequences (x_n), (y_n) in $[a, b]$ satisfying:

$$|x_n - y_n| < \frac{1}{n} \quad \text{and} \quad |f(x_n) - f(y_n)| \geq \varepsilon \quad \text{for all} \quad n \in \mathbb{N}$$

Now the sequence $(x_n) \subset [a, b]$ is bounded, so that by the Bolzano–Weierstrass Theorem some subsequence $(x_{n_r})_{r \geq 1}$ converges to a point $c \in [a, b]$. But as $|x_{n_r} - y_{n_r}| < \frac{1}{n_r}$ for all $r \geq 1$, it follows that $(y_{n_r})_{r \geq 1}$ also converges to c. Since f is continuous at c by hypothesis, we have:

$$\lim_{r \to \infty} f(x_{n_r}) = f(c) = \lim_{r \to \infty} f(y_{n_r})$$

(Note that if c is an endpoint of $[a, b]$ we can use left- (resp. right-) continuity to deduce the same.) But now $\lim_{n \to \infty}(f(x_{n_r}) - f(y_{n_r})) = 0$, and this contradicts the fact that the inequality $|f(x_{n_r}) - f(y_{n_r})| \geq \varepsilon$ is assumed to hold for all $r \geq 1$. Thus f is uniformly continuous on $[a, b]$.

⊕ *Example 10*

The function $f(x) = x^2$ is *not* uniformly continuous on $\mathbb{R}$. Take $\varepsilon = 1$ and consider $x = n + \frac{1}{n}$ and $y = n$, then we obtain $f(x) - f(y) = (n + \frac{1}{n})^2 - n^2 = 2 + \frac{1}{n^2}$ for all n, even though the distance $x - y = \frac{1}{n}$ can be made smaller than any given $\delta > 0$.

EXERCISE ON 7.4

1. Show that a bounded continuous function can fail to be uniformly continuous on a bounded interval, by considering $f(x) = \sin(\frac{1}{x})$ on $(0,1)$. Why does this not contradict Theorem 4?

A detective story: what did Cauchy mean?

We shall use Theorem 4 to good effect when discussing the Riemann integral in Chapter 10. If you feel that the difference between these two concepts of convergence is rather subtle, it may be some consolation that even the great Augustin-Louis Cauchy, who was renowned for his care and precision, had some difficulty in distinguishing between them! Cauchy's famous text *Cours d'Analyse* (A Course in Analysis) was published in 1821 and had a profound effect on subsequent authors; so much so that many still regard him as the 'Father of

Analysis'. Since he was the first to establish Analysis as a study in its own right, separate from Geometry and Algebra, this seems an entirely reasonable claim at first sight. His contribution includes the first consistent attempt to base the theory of functions (including the ideas of continuity and convergence, and their role in the Calculus) on a single concept, that of 'variable', by which he understood a sequence of 'values, taken one after another'.

As yet, no-one had given a consistent definition of the concept of 'real number' on which we have based our definitions in this book. Cauchy was not really explicit about this either, but his 'sequential' notion of variable enabled him to introduce many of the ideas and theorems we have encountered so far – we have, after all, also taken the convergence of *sequences* as our fundamental idea. Cauchy, however, was still firmly wedded to the concept of 'infinitesimal', by which he meant the 'eventual' values taken by a variable which is convergent to 0. Since he did not give a precise definition of his number system, however, it is not clear in what sense the infinitesimals should be regarded as fixed or 'moving' quantities (and they clearly can't be both!). This ultimately led him to make claims which were difficult to justify without further explanation – and that explanation (provided by Weierstrass and his co-workers some 25 years later) is exactly what led to the definition of *uniform* notions of continuity and convergence.

We can illustrate this with Cauchy's most famous 'mistake'. He claimed in the '*Cours*' that series of continuous functions will always lead to continuous sums, or, rather more precisely:

• *Cauchy's Sum Theorem*

'If the different terms of the series $u_0 + u_1 + \ldots + u_n + \ldots$ are functions of some variable x, continuous with respect to this variable in the neighbourhood of a particular value for which the series is convergent, then the sum s of the series is also, in the neighbourhood of this particular value, a continuous function of x.'

As it quickly transpired, if we read the claim literally, using *our* definition of continuity (and convergence) then it is slightly 'too good to be true'. If fact, the Norwegian mathematician Niels Abel showed as early as 1826 that taking $u_n(x) = (-1)^{n-1} \frac{\sin x}{n}$ yields a simple counter-example $s(x) = \sum_{n=1}^{\infty} u_n(x)$ which is not continuous at $x = \pi$. (This is an example of a *Fourier sine series*.)

Cauchy, however, clung to his Theorem and its proof for 20 years. Examining the proof quickly shows the difficulty – the problem for subsequent historians has been to *interpret* the proof to try to tease out what Cauchy actually meant by it!

Cauchy argued quite simply: let $s_n(x) = \sum_{k=1}^{n} u(x)$ be the nth partial sum of the series, and write $s(x) = s_n(x) + r_n(x)$, so that $r_n(x) = \sum_{k=n+1}^{\infty} u_k(x)$ for each x. Now let α be an 'infinitesimal increment' and compare $s(x + \alpha)$ with $s(x)$. We need to show that their difference is also infinitesimal. But

$$|s(x + \alpha) - s(x)| \leq |s_n(x + \alpha)| + |r_n(x + \alpha)| + |r_n(x)|$$

The first term on the right is infinitesimal, since the sum of *finitely* many continuous functions is continuous, and Cauchy claims that the last two terms become infinitesimal 'at the same time if one attributes to n a very considerable value'. The result follows.

The problem is, of course, that we cannot necessarily use the same n, irrespective of the choice of the 'increment' α. On the other hand, if the sums (s_n) are *uniformly* continuous on an interval surrounding x (in Cauchy's terms, a 'neighbourhood of x') then the proof can be rescued with a little care.

Cauchy's 'mistake' has given historians much to chew on in trying to interpret his definitions and the development of the ideas of Analysis which followed the *Cours*. The moral of our little glimpse of these controversies is to show how crucial it is to be as clear as possible about the *nature* of the objects we are dealing with – it is this precision which gives the *axiomatic approach* its power; even though, as we have seen, the inventors of these ideas are not always clear about what they mean!

Summary

In this chapter the consequences of the completeness of $\mathbb{R}$ were explored in the context of continuous real functions. In contrast to Chapter 6, where the definition of continuity was tested on sometimes complicated examples to illustrate that our intuitive picture will not always suffice, here we have restricted our functions to interval domains and found that the 'obvious' properties which that picture suggests are indeed true in this setting. The proofs, however, rely strongly on earlier results which use the completeness of $\mathbb{R}$.

The applications of the *IVT*, *Inverse*, and *Boundedness Theorems* were seen to be very powerful in helping to find solutions of various equations. The iterative procedures which we began to discuss in Chapter 2 can now be justified more fully, and their convergence guaranteed under quite mild assumptions. The above theorems impose strong limitations on the shape and complexity of continuous functions which have specified interval domains. In particular, we introduced the concept of *uniform* continuity and showed that for closed bounded interval domains, this stronger idea coincides with our earlier definition of continuity. This fact will prove very useful in later work.

FURTHER EXERCISES

1. For the following pairs of sets A, B *either* find a continuous function $f : A \longmapsto B$, giving its formula and sketching the graph, *or* explain why no such function exists.
   (i) $A = \mathbb{R}$, $B = [0, \infty)$
   (ii) $A = (0, \infty)$, $B = \mathbb{R}$
   (iii) $A = [0, 1]$, $B = [0, \infty)$
   (iv) $A = [0, \infty)$, $B = \mathbb{Q}$

2. Give an example of a function f whose domain and range are the interval $[0, 1]$, but such that f is discontinuous at each point.

3. Use the continuity of $x \longmapsto x^{\frac{1}{n}}$ to find $\lim_{x \to 1} \frac{1-x}{1-x^r}$ when $r = \frac{m}{n} \in \mathbb{Q}$.

4. Prove that if f is uniformly continuous on a bounded interval, then f is bounded. Does this remain true if:
   (i) f is continuous, but not uniformly continuous?
   (ii) the interval is unbounded?

8 • Differentiable Real Functions

We are ready, at last, to turn to the Calculus! The inventors of the Calculus did not, of course, have the elaborate machinery of Analysis behind them when they first developed its techniques, and even today students first learn to apply these techniques before they embark on the theorems which underpin them. Thus, for example, while the companion text *Calculus and ODEs* by David Pearson gives a clear intuitive idea of what is meant by ideas such as *limit*, *continuity* and *derivative*, no formal definitions are given at that stage – to do so would require fuller familiarity with the real number system, as we have seen. Having now worked through the impact of the formal definition of limit and continuity, we are ready to tackle the main applications: to discuss, with precision, the properties of derivatives and integrals of real functions.

8.1 Tangents: prime and ultimate ratios

Nonetheless we shall quickly recap the basic idea which leads to the differential calculus: the concept of an *instantaneous rate of change*. If we use the graph of a real function f to represent a physical quantity, such as the trajectory of a moving object (a football, rocket, or bird in flight, for example), then typically the graph will not be a straight line, since the motion of the object changes as a result of forces (such as gravity and friction) acting on it. To discuss the *variability* of motion – other than linear motion, where the rate of change is constant, i.e. remains the same throughout – we therefore need a *local* concept of the variation of f.

The problem of turning this vague statement into usable mathematics largely defeated the Ancient Greek philosophers and mathematicians, since their thinking was bound up with the geometry of uniformly regular figures. Their dynamics never really developed a widely applicable concept of variable motion; in part this was due to the influence of Aristotle, who insisted that if a body has a given velocity at some time, then that velocity has to *persist* for a length of time, so that the notion of 'instantaneous velocity' seemed paradoxical.

As ever, mathematicians overcame this seemingly impassable hurdle by *classifying* different types of motion, and studying each in turn, from the simplest to the most complex. The Mediaeval Scholastic philosophers argued that all motion is in one of three types:

(i) *uniform motion*, which occurs at a fixed velocity; so that the velocity 'graph' v is a horizontal line (with the area $ABDC$ in Figure 8.1(i) representing the displacement!)

(ii) *uniformly difform motion*, which proceeds at a constant acceleration (or as they would have it, a constant rate of change in 'intensity'). Here v is shown as the line *AC* and the displacement is represented by the area of the triangle *ABC* in Figure 8.1(ii). The famous *Merton Rule* (claimed to have been stated at Merton College, Oxford, around 1328) then pointed out this displacement is equal to that produced when travelling at a constant velocity equal to the average of the initial and final velocities – i.e. the area of triangle *ADC* is equal to that of rectangle *BCDE*.

(iii) the third form of motion, which they naturally called *difformly difform motion*, had none of these nice qualities, since *v* is not now given by a straight line, and thus needed different techniques for its study. It seemed natural to *approximate* the non-constant rates of change, i.e. non-linear curves such as *PRQ* (as in Figure 8.1(iii)), by chords *PQ*, and to see what happens when the length of the chord approaches 0. This leads one to consider the *slope of the tangent* to the curve at *P* as a representation of the 'instantaneous' velocity at *P*.

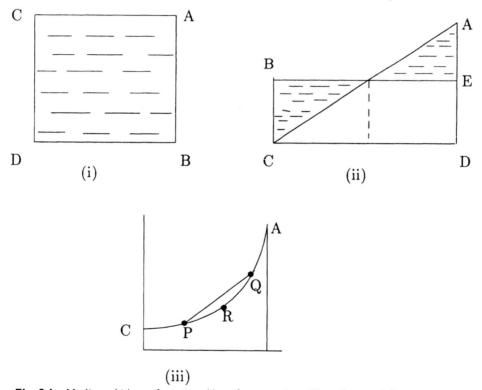

Fig. 8.1 Mediaeval ideas of motion: (i) uniform motion, (ii) uniformly difform motion, (iii) difformly difform motion

The process of turning these ideas into a *formalism* by which we can 'read off' the slope of this tangent from the formula for the curve required several centuries of further work, but the basic idea remains the same. Of course, the problem of avoiding 'division by zero' remained a real one: while the *average* rate of change in *f* on the interval [*a*, *a* + *h*] makes perfectly good sense as

$$\frac{f(a + h) - f(a)}{h}$$

the ratio loses its meaning when $h = 0$. Without a consistent theory of limits, even Isaac Newton struggled in vain to give a precise meaning to this idea: *in practice* he worked quite readily with the 'infinitesimal increments' in the variables x, y (where $y = f(x)$ in our notation). He imagined these increments $\dot{x}, \dot{y}$ as the increase (or decrease) in the values of x and y in an 'instant' of time, so that their *ratio* $\frac{\dot{y}}{\dot{x}}$ *represented the instantaneous rate of change in y relative to x*. Thinking of the graph of $y = f(x)$ as a 'flowing' quantity or *fluent*, Newton referred to their rates of change as *fluxions*, and argued that their ratio is what we would find just as the motion *starts* away from P, or, indeed, just as it *arrives* at P. In his words, taken from *The Quadrature of Curves* published in 1693:

> Fluxions are very nearly the Augments of the Fluents, generated in equal, but infinitely small parts of Time; and to speak exactly, are in the *Prime Ratio* of the nascent Augments: ... Tis the same thing if the Fluxions be taken in the *Ultimate Ratio* of the Evanescent Parts.

Thus Newton struggles to give meaning to the derivative at P as the ratio of the change in y relative to the change in x just as the motion is 'being born' at P or 'expires' at P – you may perhaps recognize in this an early attempt to express the right- (respectively left-) hand limit of the ratio $\frac{f(a+h)-f(a)}{h}$ as $h \downarrow 0$ or $h \uparrow 0$. Fortunately, with the machinery we now have at our disposal, we can be much less convoluted in expressing these ideas!

8.2 The derivative as a limit

The real functions we consider usually have fairly 'nice' domains, such as (finite unions of) intervals, so we shall adopt a slightly restricted definition of the derivative, valid for points a in the *interior* of the domain of the real function f. A point a is said to be *interior* to A if some neighbourhood $N(a, \delta) = (a - \delta, a + \delta)$ of a is a subset of A.

We restrict our initial attention to such points because the *value $f'(a)$* of the derivative of f at a is defined as a *limit*:

$$f'(a) = \lim_{x \to a} \frac{f(x) - f(a)}{x - a}$$

For this limit to make sense, we need to know that there is a real number (which we shall denote by $f'(a)$) such that, whenever the sequence (x_n) in $\mathcal{D}_f$, with $x_n \neq a$ for each n, converges to a, then the difference $|\frac{f(x_n)-f(a)}{x_n-a} - f'(a)|$ will be arbitrarily small for sufficiently large n. Implicit in this is the *assumption* that there actually exists at least one sequence (x_n) in $\mathcal{D}_f$ which converges to a, as we discussed in Chapter 5. Taking a to be an *interior* point of $\mathcal{D}_f$ ensures that such a sequence can always be found. So we make the following:

• *Definition I*

The real function f is *differentiable at a* if for some $\delta > 0$ the domain $\mathcal{D}_f$ contains the δ-neighbourhood $N(a, \delta)$ and $f'(a) := \lim_{x \to a} \frac{f(x)-f(a)}{x-a}$ exists in $\mathbb{R}$. We say that f is *differentiable on A* if it is differentiable at each interior point of A, and we call the real function $f' : a \mapsto f'(a)$ the *derivative* of f on A.

This definition will suffice for most of the examples we shall encounter. We can easily rewrite it in the form we had earlier, namely

$$f'(a) = \lim_{h \to 0} \frac{f(a+h) - f(a)}{h}$$

since $x = a + h$ converges to a if and only if $h \to 0$. It is this form which is usually recast as $\frac{dy}{dx} = \lim_{\Delta x \to 0} \frac{\Delta y}{\Delta x}$ when one considers tangents to the curve $y = f(x)$.

Example 1

The standard 'schoolbook' example is $f(x) = x^2$, where, for any $a \in \mathbb{R}$, $\frac{f(x)-f(a)}{x-a} = \frac{x^2-a^2}{x-a} = x + a$ whenever $x \neq a$. Hence $f'(a) = \lim_{x \to a}(x + a) = 2a$ for any $a \in \mathbb{R}$. The *derivative* of f is the *function* f' with $f'(x) = 2x$ for all $x \in \mathbb{R}$, and the *value* of the derivative at $a \in \mathbb{R}$ is $f'(a) = 2a$. It is important to distinguish between these; not least because we will soon want to repeat the process, and find the derivative of f', and this makes sense only for functions, not for their values!

However, we should also make provision for 'one-sided tangents' at endpoints of interval domains. For this we can simply use one-sided limits instead:

Definition 2

The real function f is *right-differentiable* at a if, for some $\delta > 0$, the interval $[a, a + \delta)$ is contained in $\mathcal{D}_f$ and the right-limit $\lim_{x \downarrow a} \frac{f(x)-f(a)}{x-a} =: f'_+(a)$ exists in $\mathbb{R}$. The number $f'_+(a)$ is the *right-derivative* of f at a.

The *left-derivative* $f'_-(a)$ is defined similarly: if for some $\delta > 0$, the interval $[a - \delta, a) \subset \mathcal{D}_f$ and $\lim_{x \uparrow a} \frac{f(x)-f(a)}{x-a} =: f'_-(a)$ exists in $\mathbb{R}$, we say that f is *left-differentiable* at a.

It is clear that the derivative $f'(a)$ exists at a if and only if both $f'_+(a)$ and $f'_-(a)$ exist and are equal.

Example 2

However, the one-sided derivatives can both exist without being equal, as the example $f : x \mapsto |x|$ shows: at $a = 0$ we have $f'_-(0) = -1$ and $f'_+(0) = 1$. Hence f is *not* differentiable at 0, although it is differentiable at each $a \neq 0$, and in fact $f'(a) = -1$ when $a < 0$, $f'(a) = 1$ when $a > 0$. Therefore the derivative f' has domain $\mathbb{R} \setminus \{0\}$ and can be written in the form $f'(x) = \frac{|x|}{x}$ for $x \neq 0$. The function $x \mapsto \frac{|x|}{x}$ is often given the arbitrary value 0 at 0 and is then denoted by $sgn(x)$ (the *sign* function). Thus if $x > 0$, $sgn(x) = 0$ if $x = 0$, and $sgn(x) = -1$ if $x < 0$. (This is the closest we can get to giving the modulus function a 'derivative'.)

Example 3

The power function $f_r : x \mapsto x^r$, defined for positive rational $r = \frac{m}{n}$ as $x \mapsto (x^{1/n})^m$, has domain $[0, \infty)$, and we shall shortly show that it is differentiable on $(0, \infty)$ for each $r \geq 0$. However, it is right-differentiable at 0 only for $r \geq 1$: to see this, consider the ratios $\frac{x^r - 0}{x - 0} = x^{r-1}$:

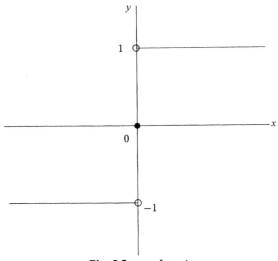

Fig. 8.2   *sgn* function

(i)   For $0 \le r < 1$ this is a negative power of x and so increases without bound as $x \downarrow 0$.

(ii)   When $r = 1$ we just have $x \mapsto x$, which has derivative 1 everywhere.

(iii)   For $r > 1$ the ratio is a positive power of x and so has limit 0 at 0.

Therefore f_r has a right-derivative at 0 if $r \ge 1$, while for $r < 1$ the right-derivative of f_r at 0 does not exist. In particular, $x \mapsto \sqrt{x}$ is not right-differentiable at 0, while $x \mapsto x\sqrt{x} = x^{3/2}$ has right-derivative 0 at 0.

EXERCISES ON 8.2

1.   Show directly from the definitions that $f(x) = \frac{1}{x}$ is differentiable at each $a \ne 0$ and find the derivative f'. Can we extend f' to include the point $a = 0$ by defining $f(0) = 0$?

2.   Explain carefully at which points the function $x \mapsto x - [x]$ is differentiable and find the value of its derivative at these points.

8.3 Differentiation and continuity

Like continuity, differentiability of a function is a *local* property: a function can be differentiable at some points without being differentiable at others, and differentiability at the point a depends only on the behaviour of f at points near a. Note also that the function $x \mapsto |x|$ is continuous at 0, but not differentiable there. Thus continuity at a does *not* imply that f is differentiable at a.

Example 4

Another (much 'wilder') example of this is given by the function $f : x \mapsto \begin{cases} x\sin\frac{1}{x} & \text{for } x \ne 0 \\ 0 & \text{for } x = 0 \end{cases}$. This is differentiable at each $x \ne 0$, since we can restrict attention to an interval about x not including 0, and apply the well-known 'rules' for combining derivatives which we shall prove shortly. Thus for each $x \neq 0$, $f'(x) = \sin\frac{1}{x} - \frac{1}{x}\cos(\frac{1}{x})$. On the other hand, $\frac{f(x)-f(0)}{x-0} = \sin\frac{1}{x}$ and this does not have a limit as $x \to 0$. Hence $f'(0)$ does not exist. Nonetheless, as we saw earlier, f is continuous at 0, since $|x\sin\frac{1}{x}| \leq |x|$.

We now show that differentiability is a strictly stronger condition than continuity:

● Theorem I ——————————————————

If the real function f is differentiable at a then it is continuous at a.

PROOF

To see this, we define a new function to represent the differential ratios, the 'chord-slope function' $\frac{f(x)-f(a)}{x-a}$. The function f has a derivative $f'(a)$ at a precisely when the extension of the chord-slope function given by

$$F_a(x) = \begin{cases} \frac{f(x)-f(a)}{x-a} & \text{if } x \neq a \\ f'(a) & \text{if } x = a \end{cases}$$

is *continuous* at a, since then

$$F_a(a) = f'(a) = \lim_{x \to a} \frac{f(x)-f(a)}{x-a} = \lim_{x \to a} F_a(x)$$

But we always have $f(x) - f(a) = F_a(x)(x - a)$ for $x \neq a$, and at $x = a$ this still holds when f is differentiable at a, since then $x - a = 0$ and $F_a(a) = f'(a)$ is finite. Thus we can write f as a combination of continuous functions: $f(x) = f(a) + F_a(x)(x - a)$ for all x. But now, in particular, f is continuous at a since both F_a and $x \mapsto x - a$ are.

This proof illustrates a useful interpretation of the derivative: since F_a is continuous at a, replacing (at least for x near a) the *variable* values $F_a(x)$ of the function F_a by the *fixed* value $F_a(a) = f'(a)$ in the expression for $f(x) = f(a) + F_a(x)(x - a)$ gives us a 'good' approximation to the graph of the function f by the straight line $y = f(a) + f'(a)(x - a)$.

Hence we have approximated f near a by an *affine* function (that is, by a straight line of the form $y = b + mx$). The *error* in making this approximation can be written as $\theta_a(x) = f(x) - y = [F_a(x) - f'(a)](x - a)$, and since $F_a(x) \to f'(a)$ as $x \to a$, this shows that the derivative $f'(a)$ gives us the *unique* affine approximation to f for which the error satisfies $\frac{\theta_a(x)}{x-a} \to 0$ as $x \to a$. (*See also* the discussion of tangents in *Calculus and ODEs*, Section 4.2.)

For a differentiable function f, therefore, the value of the derivative at a gives the slope of the *best affine* approximation to f near a.

TUTORIAL PROBLEM 8.1

Discuss carefully why this approximation is the *unique* one which satisfies the condition $\frac{\theta_a(x)}{x-a} \to 0$ as $x \to a$. In what sense does this make the approximation the 'best' affine one?

EXERCISES ON 8.3

1. Write down the precise definition of the *second derivative* f'' of a real function f at an interior point a of its domain. State carefully which functions are involved, and where they need to be defined for the definition to make sense.

2. Let $P(x) = x^7 - 3x^6 + 3x^5 - x^3 - 3x^2 + 5x - 2$. Assuming that polynomials can be differentiated term by term and that the derivative of $x \mapsto x^n$ is $x \mapsto nx^{n-1}$ for all $n \geq 1$, show that $(x - 1)^3$ is a factor of $P(x)$. (*Hint:* consider $P(1)$, $P'(1)$ and $P''(1)$.)

3. Suppose that f is differentiable at a. Show that

$$\lim_{h \downarrow 0} \frac{f(a + h) - f(a - h)}{2h} = f'(a)$$

By considering $f(x) = |x|$, show that the above limit may exist even if f is not differentiable at a.

8.4 Combining derivatives

The next 'jumbo' theorem summarizes the basic rules for combining derivatives (which you have been using for some time!). They enable us to extend the class of functions we can differentiate very quickly from modest beginnings – for example, it is obvious from the definitions that the constant function $x \mapsto c$ has derivative 0 everywhere, and that $x \mapsto x$ has derivative $x \mapsto 1$. The rules then enable us to differentiate polynomials and rational functions without further effort, and also deal with composite and inverse functions (which will include rational power functions). To deal with *transcendental* functions like exp, log, sin, cos, however, we shall still need further theorems on the differentiation of functions defined by power series.

● *Theorem 2* ─────────────────────────────

(a) Algebra of derivatives:
Suppose f and g are differentiable at a. Then $f + g, f - g, f.g$ are differentiable at a, and if $g(a) \neq 0$, $\frac{f}{g}$ is differentiable at a. The following rules for differentiation apply:
  (i) $(f + g)'(a) = f'(a) + g'(a)$; $(f - g)'(a) = f'(a) - g'(a)$;
  (ii) $(f.g)'(a) = f'(a).g(a) + f(a).g'(a)$;
  (iii) $\left(\frac{f}{g}\right)'(a) = \frac{f'(a)g(a) - g'(a)f(a)}{(g(a))^2}$.
   Similar rules apply to one-sided derivatives.

(b) Chain rule
If f is differentiable at a and g is differentiable at $b = f(a)$ then their composition $g \circ f$ is differentiable at a and $(g \circ f)'(a) = g'(b).f'(a)$.
   (Informally: $\frac{dz}{dx} = \frac{dz}{dy} \cdot \frac{dy}{dx}$)

(c) Inverse Function Theorem
If f is continuous and one-one on the open interval (a, b) and differentiable at $c \in (a, b)$, with $f'(c) \neq 0$, then the inverse function f^{-1} is differentiable at $d = f(c)$, and its derivative is given by $(f^{-1})'(d) = \frac{1}{f'(c)}$.

(Informally: $\frac{dx}{dy} = \frac{1}{dy/dx}$)

PROOF

(a) If $x \neq a$,

$$\frac{(f+g)(x) - (f+g)(a)}{x-a} = \frac{f(x) - f(a)}{x-a} + \frac{g(x) - g(a)}{x-a}$$

and similarly for $(f - g)$. Taking limits as $x \to a$ on each side yields (i).

For the product rule we must work a little harder: we write

$$(f \cdot g)(x) - (f \cdot g)(a) = [f(x)g(x) - f(a)g(x)] + [f(a)g(x) - f(a)g(a)]$$

so that for $x \neq a$:

$$\frac{(f \cdot g)(x) - (f \cdot g)(a)}{x-a} = g(x)\frac{f(x) - f(a)}{x-a} + f(a)\frac{g(x) - g(a)}{x-a}$$

Taking limits on both sides is not quite as straightforward this time, since the first product on the right involves two functions of x. However, by Theorem 1 g is continuous at a, so $\lim_{x \to a} g(x) = g(a)$, and hence both functions in the product are continuous at a. Thus the limit on the right becomes $g(a)f'(a) + f(a)g'(a)$ as required.

For the quotient rule we first consider the derivative of $\frac{1}{g}$ at a: since g is continuous at a and $g(a) \neq 0$, we can restrict to a neighbourhood $N(a, \delta)$ of a on which $g(x) \neq 0$. Then we can write

$$\frac{\frac{1}{g(x)} - \frac{1}{g(a)}}{x-a} = \frac{g(a) - g(x)}{g(a)g(x)(x-a)} = -\frac{1}{g(a)g(x)}\left[\frac{g(x) - g(a)}{x-a}\right]$$

and using the continuity of g at a once more, we see that the limit on the right is $-\frac{g'(a)}{(g(a))^2}$ when $x \to a$. Now combine this with the product rule on the product $(\frac{f}{g}) = f \cdot (\frac{1}{g})$ to obtain:

$$\left(\frac{f}{g}\right)'(a) = f'(a)\left(\frac{1}{g}\right)(a) + f(a)\left(\frac{1}{g}\right)'(a)$$

which equals

$$\frac{f'(a)}{g(a)} + f(a)\left[-\frac{g'(a)}{(g(a))^2}\right] = \frac{f'(a)g(a) - f(a)g'(a)}{(g(a))^2}$$

(b) Consider the chord-slope functions for f at a and g at b:

$$F_a(x) = \begin{cases} \frac{f(x) - f(a)}{x-a} & \text{if } x \neq a \\ f'(a) & \text{if } x = a \end{cases}$$

and writing $y = f(x)$ we have

$$G_b(y) = \begin{cases} \frac{g(y) - g(b)}{y-b} & \text{if } y \neq b \\ g'(b) & \text{if } y = b \end{cases}$$

But we are given that F_a is continuous at a and G_b is continuous at $b = f(a)$. For all x, y we have:

$$f(x) = f(a) + F_a(x)(x-a) \text{ and } g(y) = g(b) + G_b(y)(y-b)$$

Now $(g \circ f)(x) = g(f(x)) = g(y) = g(b) + (y - b)G_b(y)$, and since $y = f(x)$ and $b = f(a)$ we can write the right-hand side as

$$g(f(a)) + (f(x) - f(a))G_b(f(x)) = g(f(a)) + (x - a)F_a(x)G_b(f(x))$$

In other words,

$$\frac{(g \circ f)(x) - (g \circ f)(a)}{x - a} = F_a(x)(G_b \circ f)(x)$$

But all the functions on the right are continuous at $x = a$, so that the limit of both sides exists as $x \to a$, and equals $F_a(a)G_b(f(a)) = f'(a)g'(b)$, and this is therefore the derivative of $g \circ f$ at a.

(c) Again write $y = f(x)$, then for $x \neq c$ we have $y \neq d = f(c)$, since the function f is one-one. Therefore we can write

$$\frac{f^{-1}(y) - f^{-1}(d)}{y - d} = \frac{x - c}{y - d} = \frac{1}{(\frac{y-d}{x-c})} = \frac{1}{(\frac{f(x)-f(c)}{x-c})}$$

which converges to $\frac{1}{f'(c)}$ when $x \to c$. This completes the proof.

⊕ *Example 5*

The rational power function $f_r : x \mapsto x^r$, with $r = \frac{m}{n} \in \mathbb{Q}$, is differentiable on $(0, \infty)$. To see this we repeatedly apply the different parts of Theorem 2: recall that $f_r = f_m \circ f_n^{-1}$ by definition, since $x^r = (x^{1/n})^m$. The function $g_n : x \mapsto x^n$ is differentiable and has derivative $f_n' : x \mapsto nx^{n-1}$ (using 2(a) repeatedly) and it is one-one on the open interval $(0, \infty)$, so that 2(c) applies. Finally we apply 2(b) to the composition. To calculate the derivative using these rules, again write $y = x^n$, then

$$(f_n^{-1})'(y) = \frac{1}{f_n'(x)} = \frac{1}{nx^{n-1}} = \frac{1}{n}y^{\frac{1}{n}-1}$$

since $x^{n-1} = (x^n)^{1-\frac{1}{n}} = y^{1-\frac{1}{n}}$. Hence the 'power rule for derivatives' also applies to the power $\frac{1}{n}$. Finally we use the chain rule to obtain: $f_r'(y) = f_m'(x)(f_n^{-1})'(y)$ for $y = x^n > 0$, i.e.

$$f_r'(y) = (mx^{m-1})\left(\frac{1}{n}y^{\frac{1}{n}-1}\right) = \frac{m}{n}y^{\frac{1}{n}(m-1)+\frac{1}{n}-1} = ry^{r-1}$$

In the next three examples we take for granted properties of the derivatives of the exponential and trigonometric functions, which we shall justify more precisely in the next chapter.

⊜ *Example 6*

Assuming, for now, that the exponential function $f(x) = \exp x$ is its own derivative (i.e. $f'(x) = \exp x$ for all $x \in \mathbb{R}$), we can use the Inverse Function Theorem to verify that the derivative of $\log = f^{-1}$ is the function $x \mapsto \frac{1}{x}$. Writing $y = f(x) = \exp x$ we have $x = f^{-1}(x) = \log y$, and according to the Inverse Function Theorem the derivative of f^{-1} at $b = f(a) = \exp a$ is thus given by $\frac{1}{f'(a)} = \frac{1}{\exp a} = \frac{1}{b}$. Hence log has derivative $x \mapsto \frac{1}{x}$, as claimed.

This enables us to extend the power rule to general real powers. We showed in Example 5 that it applies to rational powers. For general powers we use the alternative form of the definition, i.e. $f_a(x) = x^a = \exp(a \log x)$, which (using the chain rule) yields $f'_a(x) = x^a(\frac{a}{x}) = ax^{a-1}$ once more.

Example 7

On the interval $(-\frac{\pi}{2}, \frac{\pi}{2})$ the function sin is one-one and differentiable (of course this will also need justification later!) and its derivative (cos) is not 0. Hence the inverse function $\sin^{-1}$ must be differentiable on $(-1, 1)$ and its derivative at $y = \sin x$ is given by $\frac{1}{\cos x} = \frac{1}{\sqrt{1 - \sin^2 x}} = \frac{1}{\sqrt{1 - y^2}}$. (Note that we take the positive square root, since $\cos > 0$ on the interval!)

Example 8

The function $f : x \longmapsto \begin{cases} x^2 \sin\frac{1}{x} & \text{if } x \neq 0 \\ 0 & \text{if } x = 0 \end{cases}$ is differentiable at every $x \neq 0$, since we can restrict to an interval centred on x which excludes 0. Thus on this interval f is given by $x \longmapsto x^2 \sin\frac{1}{x}$, and its derivative is therefore the function $x \longmapsto 2x \sin\frac{1}{x} - \cos\frac{1}{x}$ throughout this interval. But f is also differentiable at 0, as we can deduce directly from the definition: $\frac{f(x) - f(0)}{x - 0} = x \sin\frac{1}{x}$ converges to 0 as $x \to 0$, as we saw earlier. Hence f' is the function with domain $\mathbb{R}$ given by $f' : x \longmapsto \begin{cases} 2x \sin\frac{1}{x} - \cos\frac{1}{x} & \text{if } x \neq 0 \\ 0 & \text{if } x = 0 \end{cases}$. However, this function is *not* continuous at 0, since, if it were, we would have exhibited $\cos\frac{1}{x}$ as the difference of two continuous functions (f' and $2x \sin\frac{1}{x}$), which is absurd. Hence the derivative can exist without being continuous.

EXERCISES ON 8.4

1. Find the derivative of $x \longmapsto x^x$ when it exists. At which point(s) is the derivative 0? Do the same for $x \longmapsto x^{(x^x)}$.
2. Where is the function $x \longmapsto \sqrt{1 - x^2} - \sqrt{x^2 - 1}$ differentiable?
3. Suppose that $f : (a, b) \longmapsto \mathbb{R}$ is one-one and differentiable, and that $f'(c) = 0$ for some $c \in (a, b)$. Use the chain rule on the function $f^{-1} \circ f$ to show that f^{-1} cannot be differentiable at $d = f(c)$.
4. Use the rules for differentiation to calculate the derivatives of $\sinh^{-1}$ and $\cosh^{-1}$.

8.5 Extreme points and curve sketching

Perhaps the most natural application of these results is to locating the extreme points (maxima and minima) of a function, and hence to sketching its graph, as the extreme points are often *turning points* of the graph of f. The following definitions help us to distinguish the different types of behaviour of the graph:

● Definition 3

The real function f has a *local maximum* at the point $a \in \mathcal{D}_f$ if there is a neighbourhood $N(a, \delta)$ of a such that $f(a) \geq f(x)$ for all $x \in \mathcal{D}_f \cap N(a, \delta)$. The maximum is *strict* if $f(a) > f(x)$ for all such x, and it is *global* (or *strict global*) if the inequalities hold throughout $\mathcal{D}_f$.

Simiiar definitions, with the inequalities reversed, are used for a (*local, global, strict, . . .*) *minimum* point. To indicate that $a \in \mathcal{D}_f$ could be either a minimum or a maximum point, we call it an *extreme point* for f. (Note that a strict global maximum must be unique, if it exists.)

The graphs of the functions $x \mapsto x^2$, $x \mapsto \sin x$, $x \mapsto 1$, $x \mapsto |x|$ illustrate some of the many possibilities. Note again that in most cases of extreme points we focus on the *local* behaviour of f; it is here where the value of the derivative provides useful information.

● *Proposition 1*

If the real function f is differentiable at a and has an extreme point at a, then $f'(a) = 0$.

(Some authors call points a where $f'(a) = 0$ *stationary points* of f.)

PROOF
Suppose that f has a local minimum at a (the other case is similar, with inequalities reversed). Then $f(x) \geq f(a)$ for all $x \in N(a, \delta) \cap \mathcal{D}_f$ for some $\delta > 0$. Since f is differentiable at a, the point a must be interior to $\mathcal{D}_f$ and so the inequality will hold throughout $N(a, \delta)$ for small enough $\delta > 0$. Hence we have $\frac{f(x)-f(a)}{x-a} \leq 0$ for $a - \delta < x < a$ and $\frac{f(x)-f(a)}{x-a} \geq 0$ for $a < x < a + \delta$. Consequently the left-derivative $f'_-(a) \leq 0$ and the right-derivative $f'_+(a) \geq 0$. Since f is differentiable at a, these two values must be equal, so that $f'(a) = 0$.

We must take care here with the *logic* of the Proposition! It does *not* claim that each point at which $f'(a) = 0$ will automatically be extreme: this would be contradicted, for example, by $f(x) = x^3$, which has $f'(x) = 3x^2$, which is 0 at $x = 0$, although 0 is not an extreme point for f, since $f(x) < 0$ whenever $x < 0$, and $f(x) > 0$ whenever $x > 0$. (This is an *example* of a 'point of inflection' for f; but that isn't quite the full story!)

Nonetheless Proposition 1 is very useful in *ruling out* points from being extreme: if f is differentiable at a and $f'(a) \neq 0$ then a *cannot* be an extreme point for f (since it would contradict the Proposition). So the extreme points come from two classes: they will either be *among* the stationary points *or* they will be points at which f fails to be differentiable. Points which fall into one of these two classes will be called *critical points*. The following very simple example contains both classes: let $f(x) = (|x| - 1)^2$, which has domain $\mathbb{R}$, and is differentiable except at $x = 0$. For $x \neq 0$, $x \mapsto |x|$ has derivative $sgn(x) = \frac{|x|}{x}$ so by the chain rule we obtain $f'(x) = 2(|x| - 1)\frac{|x|}{x}$, which is 0 only if $x = 1$ or -1. Hence the three *possible* extreme points are -1, 0 and 1. But $f(x) \geq 0$ for all x, and $f(-1) = f(1) = 0$, so both of these are (strict local, and also global) minimum points, while there is a strict local maximum at $x = 0$ (since $f(x) < 1 = f(0)$ for all $x \in N(0, 2)$, for example). This maximum is not global, since (e.g.) $f(3) > 1$. It is now easy to sketch the graph of f.

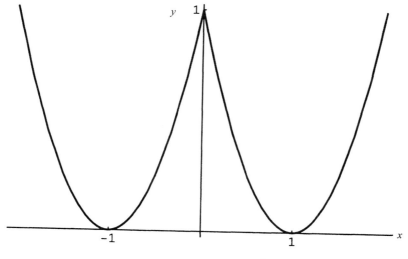

Fig. 8.3 $x \mapsto (|x| - 1)^2$

In this example we were able to deduce what happens 'near' the critical points simply by inspection. But we can use the sign of f' to gain information about the behaviour of f in a neighbourhood of a: for example, if $f'(a) > 0$, then the *continuity* of the chord-slope function F at a tells us that the ratios $\frac{f(x)-f(a)}{x-a}$ must be positive for $x \in N(a, \delta)$, provided $\delta > 0$ is sufficiently small. Hence we must have $f(x) < f(a)$ for $a - \delta < x < a$ and $f(x) > f(a)$ for $a < x < a + \delta$ (since the sign of $x - a$ changes as x increases through these intervals). Thus: f is *strictly increasing* throughout some neighbourhood $N(a, \delta)$ if $f'(a) > 0$. Our next task will be to extend this idea when $f' > 0$ *throughout* any given set, i.e. we want to examine what 'global' deductions we can make from the properties of the derivative. As with continuous functions, we need the set in question to be an interval for these results. We shall tackle these questions in the next chapter.

EXERCISES ON 8.5

1. Decide at which points the function $x \mapsto |x^2(x - 1)|$ is differentiable and find its derivative at each such point.
2. Given that $(2, -1)$ is a stationary point of the function $f(x) = \frac{ax+b}{x^2-5x+4}$, find the values of a and b.
3. Show that $x \mapsto \frac{\sin(x+a)}{\sin(x+b)}$ has no extreme points, whatever the values of a and b. Sketch the graph of this function when $a = 1$ and $b = -1$.

Summary

The material covered in this chapter is probably more familiar to you than that of the preceding ones! But in order to provide rigorous foundations for differentiation we need to have a proper grasp of the ideas of limit and continuity, and these have been applied consistently throughout, both in justifying the definitions and to prove the basic theorems which provide the algorithms used in the Calculus. Differentiable functions were shown to be continuous, but the converse does not hold. One-sided derivatives proved useful, in much the same way as did one-sided limits earlier, in handling a number of examples. The rules for combining derivatives provided us with tools for analysing general power functions, though discussion of the derivatives of functions defined by power series have been deferred to the next chapter. Finally, we distinguished between extreme, critical and stationary points and showed how the local properties of the derivative enable us to find extreme points and sketch graphs.

FURTHER EXERCISES

1. Prove the following *sandwich principle* for differentiation:
   If the functions f, g, h satisfy $g(x) \leq f(x) \leq h(x)$ for all $x \in N(c, \delta)$, while $g(c) = f(c) = h(c)$, and if g and h are differentiable at c then so is f and all three derivatives coincide at c.

2. Show that $f(x) = \begin{cases} x^2 \cos(\frac{1}{x}) & x \neq 0 \\ 0 & x=0 \end{cases}$ is differentiable at 0. Is the derivative continuous at 0? Explain.

3. Suppose that f is defined throughout $N(a, \delta)$ for some $\delta > 0$ and its second derivative f'' exists at a. Show that

   $$\lim_{h \to 0} \frac{f(a+h) + f(a-h) - 2f(a)}{h^2} = f''(x)$$

   Give an example where this limit exists, but f is not twice differentiable at a.

9 • Mean Values and Taylor Series

We have seen that a constant function has zero derivative at each point. But is the converse true? It certainly seems plausible from our picture of the derivative as the slope of the tangent to a curve: when the tangent is a horizontal line *everywhere*, then so should the function be! Similarly, we saw that the *local* behaviour of the function can be read off from the values of its derivative: for example, it is easy to believe that a function is strictly increasing on an interval I precisely when its derivative is positive at each interior point of I.

It is time to prove the basic theorems which make these ideas more precise. The principal result is the *Mean Value Theorem (MVT)*, which ensures (for 'nice' curves!) that for a given chord PQ joining two points on the curve f, we can always find a point c 'in between' at which the tangent to the curve is parallel to PQ. But if the curve increases from P to Q, then the slope of the chord PQ measures the average or *mean* increase in the function over the given interval. Hence the slope $f'(c)$ of the tangent provides the principal coefficient for a *straight line approximation* to f. Repeating the process again and again, we obtain approximations to f by *polynomials* of ever higher degree: the *Taylor polynomials* of f.

9.1 The Mean Value Theorem

The key to analysing the global behaviour of functions via the derivative lies in the *completeness* property of the real line, which allows us to use the Boundedness Theorem of Chapter 7. From this we can conclude that if the differentiable real function f takes the same value at two points a and b, then the tangent to f will be horizontal at some point between a and b. This is the idea behind *Rolle's Theorem*, named after the French mathematician Michel Rolle (1652–1719) who announced the result in 1691 in a book which was not widely read or understood at the time. Ironically, Rolle started out as a fierce critic of the techniques of Newton, Leibniz and their followers – yet today his fame rests entirely on a theorem which is firmly embedded in the Calculus.

• Theorem 1—Rolle's Theorem ────────────

If f is continuous on $[a, b]$ and differentiable on (a, b), with $f(a) = f(b)$, then there is a point $c \in (a, b)$ with $f'(c) = 0$.

PROOF

If f is constant on $[a, b]$ then $f'(c) = 0$ for *every* $c \in (a, b)$. Otherwise f takes at least two values in $[a, b]$. But f is continuous on the closed and bounded interval $[a, b]$, so it attains both its sup and its inf on $[a, b]$ at points of $[a, b]$. The extreme values cannot both be taken at the endpoints of the interval, since $f(a) = f(b)$. Hence there is at least one extreme point c for f contained in the open interval (a, b). Since f is differentiable at c, Proposition 1 in Chapter 8 ensures that $f'(c) = 0$.

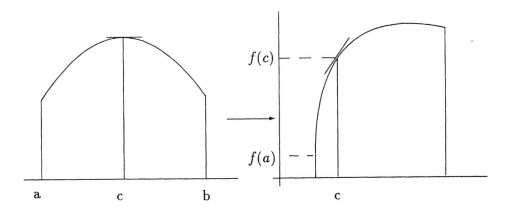

Fig. 9.1. Rolle's Theorem and the *MVT*

Note that we did not need f to be *differentiable* at the endpoints a, b of the interval, but we did need it to be *continuous* throughout the closed, bounded interval $[a, b]$, in order to apply the Boundedness Theorem. By turning the 'picture' which illustrates Rolle's Theorem through an angle we also obtain the (only apparently more general) Mean Value Theorem, which is the key result of this section.

● *Theorem 2—The Mean Value Theorem (MVT)* ————

If f is continuous on $[a, b]$ and differentiable on (a, b) then there is a point $c \in (a, b)$ such that $f'(c) = \frac{f(b)-f(a)}{b-a}$.

PROOF
The proof is now a simple piece of algebra: let $g(x) = f(x) - \mu(x - a)$, where $\mu = \frac{f(b)-f(a)}{b-a}$. Then g satisfies the same conditions as f, that is, g is differentiable on (a, b) and continuous on $[a, b]$. Moreover,

$$g(b) = f(b) - \mu(b - a) = f(b) - \frac{f(b) - f(a)}{b - a}(b - a) = f(a)$$

while $g(a) = f(a) - \mu(a - a) = f(a)$. Thus $g(b) = g(a)$, and therefore we can apply Rolle's Theorem to g to find $c \in (a, b)$ such that $g'(c) = 0$. But $g'(x) = f'(x) - \mu.1$, for all $x \in (a, b)$, so that $f'(c) = \mu = \frac{f(b)-f(a)}{b-a}$ as required.

These results have many far-reaching applications. Among the most immediate is the purpose for which Rolle originally proved his theorem, namely to find estimates for roots to equations. Our first examples demonstrate the technique to estimate roots of polynomials and approximate values of trigonometric functions.

Example 1

To find an estimate for $(80)^{\frac{1}{4}}$ we consider the function $f(x) = x^{\frac{1}{4}}$ on the interval $[80, 81]$, so that by the *MVT*

$$\frac{f(81) - f(80)}{81 - 80} = f'(c)$$

for some $c \in (80, 81)$. We chose this interval since $f(81) = 3$, so that we have $3 - f(80) = \frac{1}{4} c^{-\frac{3}{4}}$ for the above c. The remaining problem is to estimate $c^{-\frac{3}{4}}$. Since c lies between 16 and 81, $c^{\frac{1}{4}}$ lies between 2 and 3, i.e. $c^{-\frac{3}{4}} \in (\frac{1}{27}, \frac{1}{8})$. Thus $3 - \frac{1}{32} < (80)^{\frac{1}{4}} < 3 - \frac{1}{108}$. This gives a rough-and-ready estimate, without much effort (or calculators!). Slightly more use of multiplication tables suggests a better estimate still: noticing that $c^{\frac{1}{4}} \in [\frac{5}{2}, 3]$ (since $\frac{625}{16} < 80$) we have $c^{-\frac{3}{4}}$ between $\frac{1}{27}$ and $\frac{8}{125}$, hence we obtain the estimate $3 - \frac{2}{125} < (80)^{\frac{1}{4}} < 3 - \frac{1}{108}$, which is roughly twice as good.

Example 2

Similarly we can use $g(x) = \tan x$ on the interval $[\frac{\pi}{5}, \frac{\pi}{4}]$: recall that $g'(x) = \sec^2 x$, so that by the *MVT*:

$$\frac{\tan(\frac{\pi}{4}) - \tan(\frac{\pi}{5})}{\frac{\pi}{4} - \frac{\pi}{5}} = \sec^2 c$$

for some c in the open interval $(\frac{\pi}{5}, \frac{\pi}{4})$.

But $\tan(\frac{\pi}{4}) = 1$, so $1 - \tan(\frac{\pi}{5}) = (\frac{\pi}{20}) \sec^2 c$. On the interval $(\frac{\pi}{6}, \frac{\pi}{4})$ the function cos decreases from $\frac{\sqrt{3}}{2}$ to $\frac{1}{\sqrt{2}}$ so $\sec^2$ increases from $\frac{4}{3}$ to 2. Hence $1 - \frac{\pi}{10} < \tan(\frac{\pi}{5}) < 1 - \frac{\pi}{15}$.

The second area of applications should be well known to students of texts such as *Calculus and ODEs*, but we are now well placed to *prove* why the methods used there to sketch curves work so well. Again it is the *MVT* which provides the key. In this section we shall merely answer the questions raised at the beginning of this chapter; consideration of the location of *extreme points* is left for the next section.

● Proposition 1

Suppose f is continuous on $[a, b]$ and differentiable on (a, b). Then we have the following:

(i) if $f' > 0$ throughout (a, b) then f is strictly increasing on $[a, b]$
(ii) if $f' \geq 0$ throughout (a, b) then f is increasing on $[a, b]$
(iii) if $f' < 0$ throughout (a, b) then f is strictly decreasing on $[a, b]$
(iv) if $f' \leq 0$ throughout (a, b) then f is decreasing on $[a, b]$
(v) if $f' = 0$ throughout (a, b) then f is constant on $[a, b]$.

PROOF
The first four cases, i.e. parts (i) to (iv), are all similar, so we shall only consider the case $f' > 0$ on (a, b) here, leaving the others as Exercises.

To see that f is strictly increasing on $[a, b]$ suppose we are given $x < y$ in $[a, b]$. Apply the *MVT* to f on the interval $[x, y]$: we find $c \in (x, y)$ such that $\frac{f(y)-f(x)}{y-x} = f'(c)$. But $f'(c) > 0$ since $c \in (a, b)$. Hence as $y - x > 0$ it follows that $f(y) - f(x) > 0$ also. Hence $x < y$ implies $f(x) < f(y)$, so f is strictly increasing on $[a, b]$.

Part (v) follows from (ii) and (iv): if $f' = 0$ then both $f' \geq 0$ and $f' \leq 0$ hold, so f is both increasing and decreasing, i.e. constant, on $[a, b]$.

(It might be more accurate to say 'non-decreasing' instead of 'increasing' and vice versa, but most authors use the terminology as we have done.)

Observe that to use this Proposition we do not need to know that f is differentiable at the endpoints of the interval $[a, b]$; nevertheless we are able to describe what happens to f throughout the whole interval. We shall exploit this new freedom in the next section to identify extreme points of f.

Note further that (v) above already solves the first problem we posed: if I is an *interval* and $f : I \mapsto I$ has $f'(x) = 0$ for all $x \in I$, then f is constant on I.

Example 3

We claim that $\exp(x) > 1 + x$ for all $x \neq 0$ in $\mathbb{R}$. This is less obvious than it may seem, since it includes the case when $x < 0$. Recall that $\exp(x) = \sum_{n=0}^{\infty} \frac{x^n}{n!}$ by definition. Thus we certainly have $\exp(x) > 1$ whenever $x > 0$. Thus given $x < y$, we have $\exp(y - x) > 1$, and so

$$\exp(y) = \exp(x + (y - x)) = \exp(x)\exp(y - x) > \exp(x)$$

Hence exp is strictly increasing on $\mathbb{R}$.

Now consider the function $h(x) = \exp(x) - 1 - x$, which has domain $\mathbb{R}$. Assuming (still!) that exp is its own derivative, we see that $h'(x) = \exp(x) - 1 > 0$ whenever $x > 0$. Thus by Proposition 1 h is strictly increasing on $[0, \infty)$. But $\exp(-x) = \frac{1}{\exp(x)}$, so that $\exp(x) < 1$ for $x < 0$, and hence $h'(x) < 0$ on $(-\infty, 0)$. Thus by the Proposition again, h is strictly decreasing on $(-\infty, 0]$. Thus 0 is the unique *(strict)* minimum point for h, and $h(0) = 0$. Hence $h(x) > 0$ whenever $x \neq 0$, which proves our claim.

EXERCISES ON 9.1

1. Complete the proof of Proposition 1.
2. Show that if $p > 1$ and $x \in (0, 1)$ then $1 - x^p < p(1 - x)$.
3. Prove the following '*IVT*' for derivatives: If f is differentiable on $[a, b]$ and $f'(a) < \gamma < f'(b)$ then there exists $c \in (a, b)$ such that $f'(c) = \gamma$. (*Hint*: apply Rolle's Theorem to $g(x) = f(x) - \gamma x$ and show that g takes its minimum value at a point $c \in (a, b)$.)

9.2 Tests for extreme points

The familiar 'Second Derivative Test' for maximum and minimum values of a real function f has the useful feature that we need only differentiate f twice at a given point c where $f'(c) = 0$. If the value $f''(c)$ is positive, then we have a minimum at c, while if $f''(c) < 0$ it is a maximum.

The fly in the ointment is the possibility that $f''(c) = 0$. Traditional texts often refer to this as yielding a 'point of inflection', but this is not quite correct. For example, if $f(x) = x^4$ then $f''(0) = 0$, yet 0 is quite obviously a *minimum* point instead! A more precise definition for c to be a *point of inflection* requires f'' to *change sign* as we move through c; that is, its graph crosses the x-axis there.

Thus we still need to develop a test for dealing with possible maximum or minimum points which can occur even when f'' is 0. But f'' may not even exist at the point, and it is in any case often tedious to differentiate a second time, especially for complicated ratios of polynomials, etc. So finding a test which avoids the need for this differentiation has some practical value. We shall see that such a test can be found very simply; and that it has wider application than the second derivative test.

In fact, the results of the previous section provide all the tools we need:

● *Proposition 2*

Suppose that f is continuous at a.
 (i) If there is a neighbourhood $N(a, \delta)$ such that $f'(x) < 0$ whenever $a - \delta < x < a$ and $f'(x) > 0$ whenever $a < x < a + \delta$, then f has a strict local minimum at a.
 (ii) If there is a neighbourhood $N(a, \delta)$ such that $f'(x) > 0$ whenever $a - \delta < x < a$ and $f'(x) < 0$ whenever $a < x < a + \delta$, then f has a strict local maximum at a.

If the above inequalities are not strict, we can still conclude the existence of a local minimum or maximum at a, but this will not be strict in general.

PROOF

We simply apply the relevant parts of Proposition 1 to each of the intervals $I_1 = [a - \frac{1}{2}\delta, a]$ and $I_2 = [a, a + \frac{1}{2}\delta]$ separately – we use these smaller intervals to ensure that f is continuous throughout the *closed* interval in question, so that we can apply Proposition 1 without difficulty. (Note that since f is differentiable at each $x \neq a$ in I_1 or I_2 it is also continuous there.) In the first case we conclude that f is strictly decreasing on I_1 and strictly increasing on I_2, and therefore f has a minimum point at a, since this is the point in which the two intervals intersect. The second case is exactly the same, with the inequalities reversed. The final comment follows similarly from parts (iii) and (iv) of Proposition 1.

⟐ *Example 4*

The function $f(x) = |x^2(x - 1)| = x^2|x - 1|$ has derivative defined by

$$f'(x) = 2x.|x - 1| + x^2.\frac{|x - 1|}{x - 1}.1$$

unless $x = 1$. Now for $x \neq 1$, $f'(x) = 0$ gives $0 = 2x + \frac{x^2}{x-1}$, i.e. $x^2 = 2x(1 - x)$. This equation has the solution $x = 0$; for $x \neq 0$ it reduces to $x = 2(1 - x)$, which is satisfied when $x = \frac{2}{3}$.

Thus 0 and $\frac{2}{3}$ are the stationary points for f, and the set of critical points is therefore simply $\{0, \frac{2}{3}, 1\}$. In this example all three are extreme points, since $f'(x)$ is negative for $x < 0$, positive on $(0, \frac{2}{3})$, negative on $(\frac{2}{3}, 1)$ and positive for $x > 1$, as we can easily check. Thus 0 and 1 are (strict local) minimum points, and $\frac{2}{3}$ is a (strict local) maximum point. Since $f \geq 0$ throughout its domain, it follows that 0 and 1 are global minimum points (as $f(0) = f(1) = 0$), but the maximum at $x = \frac{2}{3}$ is not global, since $f(x)$ increases without bound as $x \to \infty$. The sketch of f is now easy to complete from this information:

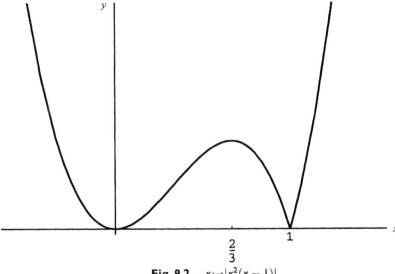

Fig. 9.2. $x \mapsto |x^2(x-1)|$

Example 5

The *Second Derivative Test* follows easily from Proposition 2: recall that we say that f is *twice differentiable* at a if the function f' is itself differentiable at a, and we denote its derivative by f''.

Now suppose that $f''(a) > 0$ and $f'(a) = 0$. To see that f then has a (strict local) minimum at a, apply the Proposition to the function $g = f'$. Since g is differentiable at a, it must be continuous there. We are given that $g(a) = 0$, but $g'(a) > 0$. Now $g'(a) = \lim_{x \to a} \frac{g(x) - g(a)}{x - a}$ by definition, and $g(a) = 0$. So the continuous function $x \mapsto \frac{g(x)}{x - a}$ will be positive throughout some neighbourhood $N(a, \delta)$ (as its limit at a is positive!) and so $g(x) < 0$ on $(a - \delta, a)$ and $g(x) > 0$ on $(a, a + \delta)$. This means that $g = f'$ satisfies the conditions of the Proposition hence f has a strict local minimum at a. The other case is similar, with inequalities reversed.

Example 6

Knowledge of the derivative also enables us to prove various inequalities for functions. As a simple example, consider the *Bernoulli Inequality* again. A quick proof of this inequality goes as follows: for $n \in \mathbb{N}$ we set $f(x) = (1 + x)^n - 1 - nx$, so that $f'(x) = n(1 + x)^{n-1} - n$. Then for $n > 1$ we have $f'(x) < 0$ for $-1 < x < 0$, since $1 + x$ is less than 1 for such x, and $f'(x) > 0$ for $x > 0$. Thus 0 is a minimum point for f on the interval $(-1, \infty)$. Now $f(0) = 0$, and so $f(x) \geq 0$ on $[-1, \infty)$. But this just says that $(1 + x)^n \geq 1 + nx$ for all $x \geq -1$ and $n \in \mathbb{N}$. Moreover, for $n > 1$ the minimum at 0 is strict, so that $(1 + x)^n = 1 + nx$ holds *only* if $x = 0$. Thus our new proof provides additional information, which was not so easy to obtain from the proof by induction used in Chapter 2.

EXERCISES ON 9.2

1. Examine the maximum and minimum values of the following functions and sketch their graphs:
    (i) $f(x) = \cos x + |\sin x|$
    (ii) $f(x) = |\sin x + \cos x|$
    (iii) $f(x) = |\sin x| + |\cos x|$
2. Find the domain and range of each of the following functions, classify their extreme points and sketch their graphs:
    (i) $x \mapsto |x^2 - x^4|$
    (ii) $x \mapsto |x| \exp(-|x|)$
    (iii) $x \mapsto x^x$
3. Show that $x \mapsto \frac{ax+b}{cx+d}$ has no extreme points and sketch its graph.

9.3 L'Hôpital's Rules and the calculation of limits

A useful extension of the *MVT* leads to a very handy method – first published in the 1696 textbook *Analyse des infiniment petits* by the Marquis de l'Hôpital, but actually due to Johann Bernoulli, who had been employed to teach the Marquis the newly invented Calculus in 1692 – for calculating 'indeterminate' limits of the form $\frac{0}{0}$, or $\frac{\infty}{\infty}$, or 0^∞. The method depends on a result now attributed to Cauchy:

• *Cauchy's Mean Value Theorem*

If f and g are continuous on $[a, b]$, differentiable on (a, b) and $g'(x) \neq 0$ on (a, b), then there is $c \in (a, b)$ such that $\frac{f'(c)}{g'(c)} = \frac{f(b)-f(a)}{g(b)-g(a)}$.

Before outlining the proof of this theorem, it is instructive to note that we cannot simply deduce it from the *MVT* by dividing the ratios $\frac{f(b)-f(a)}{b-a}$ and $\frac{g(b)-g(a)}{b-a}$ by each other: although there will be points $c, d \in (a, b)$ such that $f'(c)$ equals the first ratio and $g'(d)$ equals the second, we have no way of ensuring that c and d are *the same point* as the Cauchy *MVT* asserts.

Here is a simple counter-example: let $[a, b] = [0, 1]$ and $f(x) = x^2$, $g(x) = x^3$. Then $\frac{f(1)-f(0)}{1-0} = 1 = f'(c) = 2c$ yields $c = \frac{1}{2}$, while $\frac{g(1)-g(0)}{1-0} = 1 = g'(d) = 3d^2$ has $d = \frac{1}{\sqrt{3}}$ as its only solution in $[0, 1]$.

However, we can deduce the Cauchy Mean Value Theorem from Rolle's Theorem much as before: since g' is non-zero throughout (a, b) we cannot have $g(a) = g(b)$ (by Rolle's Theorem!). Hence the ratio $\rho = \frac{f(b)-f(a)}{g(b)-g(a)}$ makes sense, and we consider the differentiable function $h(x) = f(x) - \rho g(x)$, which has $h(b) = h(a)$ by our choice of ρ. Thus we can apply Rolle's Theorem to h, and deduce that $h'(c) = 0$ for some $c \in (a, b)$. Therefore $f'(c) = \rho g'(c)$, as required.

We proceed to the statement and proof of

• *L'Hôpital's Rules*

Rule 1: If f and g both tend to 0 as $x \to a$ then

$$\lim_{x \to a} \frac{f(x)}{g(x)} = \lim_{x \to a} \frac{f'(x)}{g'(x)}$$

whenever the latter limit exists.

Rule 2: The same applies if both f and g tend to ∞ as $x \to a$.

PROOF OF THE RULES

We prove only the first rule here; the second is similar. In order to apply the Cauchy Mean Value Theorem on the interval $(a - \delta, a + \delta)$ for some $\delta > 0$, note first that the existence of the second limit above guarantees that $g'(x) \neq 0$ throughout some neighbourhood of a. Next, $\lim_{x \to a} f(x) = \lim_{x \to a} g(x) = 0$ means that f and g are continuous at a if we define $f(a) = g(a) = 0$. Hence we obtain for $x \in N(a, \delta)$:

$$\frac{f(x)}{g(x)} = \frac{f(x) - f(a)}{g(x) - g(a)} = \frac{f'(c)}{g'(c)}$$

for some c between x and a. But as $x \to a$ we must also have $c \to a$, and the limit $\lim_{x \to a} \frac{f'(x)}{g'(x)}$ exists. So this limit must equal $\lim_{c \to a} \frac{f'(c)}{g'(c)}$, hence also equals $\lim_{x \to a} \frac{f(x)}{g(x)}$, as required.

The usefulness and versatility of these Rules will be clear from the variety of examples which we can now tackle with confidence.

Example 7

(i) $\lim_{x \to 0} \frac{\sin x}{x}$ has $f(x) = \sin x$, $g(x) = x$, which both tend to 0 at 0. Hence the limit is $\lim_{x \to 0} \frac{\cos x}{1} = 1$.

(ii) We can apply the rules repeatedly, as for example in the following:

$$\lim_{x \to 0} \frac{\tan x - x}{x - \sin x} = \lim_{x \to 0} \frac{\sec^2 x - 1}{1 - \cos x} = \lim_{x \to 0} \frac{2 \sec^3 x \tan x}{\sin x}$$

$$= \lim_{x \to 0} \frac{2}{\cos^4 x} = 2$$

Here we have first equated $\lim_{x \to 0} \frac{f(x)}{g(x)}$ with $\lim_{x \to 0} \frac{f'(x)}{g'(x)}$, then applied the same to f', g' and their derivatives. At each stage we need to be careful to make sure that the functions concerned satisfy the conditions of the Rules.

(iii) $\lim_{x \to 1} x^{\frac{1}{(x-1)}}$ has the form '1^∞', which does not fit the picture directly. However, we can first take logs to consider $\frac{1}{(x-1)} \log x$. Now $\frac{\log x}{x-1}$ has $f(x) = \log x$ and $g(x) = x - 1$, both approaching 0 as x approaches 1. So $\lim_{x \to 1} \frac{\log x}{x-1} = \lim_{x \to 1} \frac{(\frac{1}{x})}{1} = 1$. Hence by continuity of exp we have $\lim_{x \to 1} x^{\frac{1}{(x-1)}} = e$.

(iv) For an example of the second rule consider $\lim_{x \downarrow 0} \frac{\log \sin x}{\log \tan x}$; since this satisfies the (one-sided version of) Rule 2, we can write the limit as $\lim_{x \downarrow 0} \frac{\cos x / \sin x}{\sec^2 x / \tan x} = 1$.

In several of the above, we have applied the Rules more than once in order to determine the limit in question. Quite generally, we can apply the above rules repeatedly as long as the functions we consider at each stage satisfy the conditions laid down. We have already written the second derivative of a real function f as $f'' = (f')'$. This can be continued by induction: the nth derivative of f is the function $f^{(n)}$ defined as the derivative of $f^{(n-1)}$ for each n. Thus L'Hôpital's Rules can be extended; for example, Rule 1 becomes:

If f and g have continuous n^{th} derivatives at a and $g^{(n)}(a) \neq 0$, while all of

$$f, g, f', g', f'', g'', \ldots, f^{(n-1)}, g^{(n-1)}$$

equal 0 at $x = a$, then

$$\lim_{x \to a} \frac{f(x)}{g(x)} = \lim_{x \to a} \frac{f^{(n)}(x)}{g^{(n)}(x)}$$

whenever the latter limit exists. This simply formalizes what we used in the examples above. Further examples can be found in the Exercises.

EXERCISES ON 9.3

1. Find the following limits:

    (i) $\lim_{x \to 0}(\frac{1}{\sin x} - \frac{1}{x})$ (ii) $\lim_{x \to 0} x \cos(\frac{1}{x})$

    (iii) $\lim_{x \to 0} \frac{1 - \cos x}{x^2}$ (iv) $\lim_{x \to 0} \frac{\log(1+tx)}{x}$ $(t \in \mathbb{R})$

    Use part (iv) to prove that $\lim_{n \to \infty}(1 + \frac{1}{n})^n = $ e. (This provides a much quicker proof of the identity $\lim_{n \to \infty} \sum_{k=0}^{\infty} \frac{1}{k!} = \lim_{n \to \infty}(1 + \frac{1}{n})^n$, which we announced in Tutorial Problem 3.7.)

2. Show that $\lim_{t \to \infty}(1 + \frac{x}{t})^t = \exp(x)$ for all $x \in \mathbb{R}$.

3. Show that if f, g are both differentiable at 0 and tend to 0 as $x \to a$, we cannot deduce the identity

    $$\lim_{x \to a} \frac{f(x)}{g(x)} = \lim_{x \to a} \frac{f'(x)}{g'(x)}$$

    from the existence of the *first* limit (although l'Hôpital's Rule allows us to claim the identity when the *second* limit exists). (*Hint:* consider $f(x) = x^2 \cos(\frac{1}{x})$ and $g(x) = x$ at $a = 0$.)

9.4 Differentiation of power series

The *MVT* provides us with the necessary tools to extend our collection of easily calculated derivatives, to include all functions defined by power series – and hence enables us at last to justify the use of 'well-known' examples such as the exponential and trigonometric functions. At the same time it furnishes the first link between the operations of differentiation and integration.

We call a real function F a *primitive* of the real function f (on the interval I) if $F' = f$ on I. Any two differentiable functions which differ by a constant will be primitives of the same function. But the converse is also true, by Proposition 1(v): if F and G both have derivative f on I, then $H = F - G$ must have $H' = 0$ on I, and so H must be constant on I. Thus any two primitives of a real function f differ by a constant. This will explain – as we shall see in Chapter 10 – why there is an 'arbitrary constant' in the 'indefinite integral' of f.

We can already make use of this fact now to *derive* a power series expansion for $\log(1 + x)$ for $|x| < 1$: for such x the geometric series with common ratio $(-x)$ converges, and $\sum_{n=0}^{\infty}(-x)^n = (1 - (-x))^{-1}$, i.e.

$$\frac{1}{1+x} = 1 - x + x^2 - x^3 + x^4 - x^5 + \dots$$

But $\log(1 + x)$ has derivative $\frac{1}{1+x}$: to see this, recall that by definition log is the inverse function of exp, so that by Example 6 following the Inverse Function Theorem in Chapter 8, the derivative of log is the function $x \mapsto \frac{1}{x}$, and hence by the Chain Rule the result follows.

Therefore $\log(1 + x)$ is a primitive of $1 - x + x^2 - x^3 + \dots$ *If we know that we can differentiate power series term by term*, then another primitive of $1 - x + x^2 - x^3 + \dots$ is clearly given by $x - \frac{x^2}{2} + \frac{x^3}{3} - \frac{x^4}{4} + \dots$. These two primitives of $\frac{1}{1+x}$ must therefore differ only by a constant, C say, i.e. $\log(1 + x) + C = x - \frac{x^2}{2} + \frac{x^3}{3} - \frac{x^4}{4} + \dots$. But if this holds whenever $|x| < 1$ it must hold for $x = 0$, and then $\log(1) + C = 0$, hence $C = 0$. We have therefore found a series representation, valid on $(-1, 1)$, for the function $f(x) = \log(1 + x)$, namely $\log(1 + x) = x - \frac{x^2}{2} + \frac{x^3}{3} - \frac{x^4}{4} + \dots$.

Similarly one could find the representation $\tan^{-1} x = x - \frac{x^3}{3} + \frac{x^5}{5} - \frac{x^7}{7} + \dots$ by starting with the geometric series $1 - x^2 + x^4 - x^6 + \dots = \frac{1}{1+x^2}$.

But in both cases we have made use of the unproven assumption that a power series *can* be differentiated term-by-term, i.e. that if $f(x) = \sum_{n=0}^{\infty} a_n x^n$ has radius of convergence R and $|x| < R$, then the derivative f' has the power series representation $f'(x) = \sum_{n=1}^{\infty} n a_n x^{n-1}$ for all such x.

A direct proof of this result can be given using Taylor's Theorem (which we shall discuss in the next section). However it follows from a more general theorem due to Weierstrass, which provides a handy sledgehammer to crack several nuts at once! To highlight the general nature of the Theorem, we state it first in terms of general series of functions, and not merely as a result about power series.

● *Theorem 3*

Let $\sum_{k\geq 1} f_k$ be a convergent series of real functions, all differentiable on the interval $I = [a, b]$. Suppose there exists a convergent series of positive real numbers $\sum_{k\geq 1} M_k$ such that $|f_k'(x)| \leq M_k$ for every $x \in I$ and all $n \geq 1$. Then, for every $x \in I$ the sum $S(x) = \sum_{k=1}^{\infty} f_k(x)$ defines a differentiable function S on I, and the derivative of S is given by $S'(x) = \sum_{k=1}^{\infty} f_k'(x)$ for each $x \in I$.

PROOF
The condition $|f_k'(x)| \leq M_k$ is what ensures that 'all is as it should be' with our series. Let $x \in I$ be fixed. Note first that the series $\sum_{k\geq 1} |f_k'(x)|$ converges absolutely by comparison with the convergent series $\sum_{k\geq 1} M_k$. Hence the series $\sum_{k\geq 1} f_k'(x)$ converges for each $x \in I$ and we need to show that its sum $\sum_{k=1}^{\infty} f_k'(x)$ satisfies the definition of the derivative $S'(x)$ of $S(x)$, i.e. that $\frac{S(y)-S(x)}{y-x}$ converges to $\sum_{k=1}^{\infty} f_k'(x)$ as $y \to x$.

Fix $\varepsilon > 0$, and choose N so large that $\sum_{k=N}^{\infty} M_k < \frac{\varepsilon}{3}$. Also, for $k = 1, 2, \ldots, N$ choose $\delta_k > 0$ so that $|\frac{f_k(x+h)-f_k(x)}{h} - f_k'(x)| < \frac{\varepsilon}{3N}$ whenever $0 < |h| < \delta_k$. (Here N can be found as the series converges, and each δ_k can be found by definition of the derivative.) Now let $0 < |h| < \delta = \min\{\delta_k : k = 1, 2, \ldots, N\}$, while also choosing h such that $x + h \in (a, b)$. For such h we have:

$$\left| \frac{S(x+h) - S(x)}{h} - \sum_{k=1}^{\infty} f_k'(x) \right| = \left| \sum_{k=1}^{\infty} \left\{ \frac{f_k(x+h) - f_k(x)}{h} - f_k'(x) \right\} \right|$$

since the series defining S is convergent on I. Now split the sum into two pieces: $\sum_{k \leq N}$ and $\sum_{k > N}$. By our choice of δ_k for each $k \leq N$ each term in the first sum is less than $\frac{\varepsilon}{3N}$, so the sum is bounded by $\frac{\varepsilon}{3}$. Each term in the second sum has the form $f_k'(c) - f_k'(x)$ for some c lying between x and $x + h$, by the *MVT* applied to f_k. Write $c = x + \theta_k h$ for some θ_k between 0 and 1. So each term in the second sum can be estimated by writing

$$|f_k'(x + \theta_k h) - f_k'(x)| \leq |f_k'(x + \theta_k h)| + |f_k'(x)| \leq 2M_k$$

since x and $x + \theta_k h$ both lie in I. Hence by our choice of N,

$$\sum_{k \geq N} |f_k'(x + \theta_k h) - f_k'(x)| \leq 2 \sum_{k \geq N} M_k < \frac{2\varepsilon}{3}$$

Adding, we find:

$$\left| \frac{S(x+h) - S(x)}{h} - \sum_{k=1}^{\infty} f_k'(x) \right| < \varepsilon$$

for $0 < |h| < \delta$. Hence $S'(x) = \sum_{k=1}^{\infty} f_k'(x)$ as required.

To apply this general result to the power series $\sum_{k \geq 0} a_k x^k$ it simply remains to find a suitable convergent series of positive numbers $\sum_k M_k$. So suppose the power series has radius of convergence R and choose $|x| \leq r < R$. The *Hadamard formula* of Chapter 4 showed that R is then also the radius of convergence of the derived series $\sum_{k \geq 1} k|a_k| r^{k-1}$. Hence the derived series converges absolutely for $r < R$, and we can take $M_k = k|a_k| r^{k-1}$, which ensures that when $|x| \leq r \leq k$ and $f_k(x) = a_k x^k$, we obtain $|f_k'(x)| \leq M_k$. Thus the condition in the Theorem is satisfied and we have proved:

● *Proposition 3*

The power series $S(x) = \sum_{k=0}^{\infty} a_k x^k$ is differentiable term-by-term on its interval of convergence, i.e. its derivative at each point of this interval is given by $S'(x) = \sum_{k=1}^{\infty} k a_k x^{k-1}$.

TUTORIAL PROBLEM 9.1

This Proposition finally provides the check that the series definitions of exp, log, sin, cos, etc. yield the familiar derivatives used in our many examples.

The claim is immediate for exp and the trigonometric functions, and we have already seen how the former gives the result for log.

Moreover, the Proposition completes our proof that $x \mapsto x^a = \exp(a \log x)$ has derivative $x \mapsto ax^{a-1}$ for $x > 0$ and *any real a*. On the other hand, $x \mapsto a^x$ (for $a > 0$) is given by $\exp(x \log a)$, and so has derivative $x \mapsto a^x \log a$ for all x.

Example 8

In Tutorial Problem 4.4 we worked hard to establish the identity $\sin^2 x + \cos^2 x = 1$, and promised that a much less arduous proof would follow later. Proposition 3 enables us to supply such a proof, since we can now verify (with our series definitions of the trigonometric functions) what the derivatives of sin and cos are. The function $f(x) = \sin^2 x + \cos^2 x$ is therefore differentiable and has derivative $f'(x) = 2 \sin x \cos x + 2 \cos x(-\sin x) = 0$ for all real x. Hence f is constant. But $f(0) = 1$, hence $f \equiv 1$, so that we have verified the identity $\sin^2 x + \cos^2 x = 1$ for all real x.

Since differentiation of power series is so simple, we would like to use it repeatedly to find derivatives of higher orders for functions defined by power series. We introduce some useful notation and terminology for this purpose.

The real function f is *n times continuously differentiable at c* if $f^{(n)}$ is continuous at c. If this holds for all $c \in (a, b)$ we say that f is *n times continuously differentiable on (a,b)*; this is written as $f \in C^n(a, b)$. If this holds for all $n \in \mathbb{N}$ we call f *infinitely differentiable on (a,b)* and write $f \in C^\infty(a, b)$. Our theorems provide many such functions: all polynomials are trivially in $C^\infty(\mathbb{R})$, as are exp, sin, cos. Since the latter functions have been defined via power series, one can ask whether *all* C^∞-functions have expressions as power series; and if not, to what extent we can *approximate* them by polynomials of arbitrarily high degree. The answers to these questions occupy the next section.

EXERCISES ON 9.4

1. Use induction to prove *Leibniz's Formula*:
   If $f, g \in C^n(\mathbb{R})$ then $h = f.g$ is in $C^n(\mathbb{R})$ and $h^{(n)}$ is given by

   $$h^{(n)}(x) = \sum_{r=0}^{n} \binom{n}{r} f^{(r)}(x) g^{(n-r)}(x)$$

   Apply the formula to $h(x) = x^4 \cos x$.
2. Use Proposition 3 to prove the identity $\cosh^2 x - \sinh^2 x = 1$.

9.5 Taylor's Theorem and series expansions

Having established how to differentiate functions defined by power series, we can iterate the process as often as we please: if the function f has power series representation $f(x) = \sum_{k=0}^{\infty} a_k x^k$ on the interval I (which we take to be centred at

0 and have radius R) then

$$f'(x) = \sum_{k=1}^{\infty} k a_k x^{k-1} = a_1 + 2a_2 x^2 + \ldots + k a_k x^{k-1} + \ldots$$

We can rewrite the last series in the form $\sum_{n=0}^{\infty} b_n x^n$ by setting $b_n = (n+1)a_{n+1}$ for $n \geq 0$. Then apply the procedure to f' instead of f, so that the derivative of f' is given by

$$f''(x) = \sum_{n=1}^{\infty} n b_n x^{n-1} = \sum_{k=2}^{\infty} k(k-1) a_k x^{k-2}$$

on I. Repeating the process m times shows that f is m times differentiable on I and the mth derivative satisfies:

$$f^{(m)}(x) = \sum_{k=m}^{\infty} k(k-1)(k-2)\ldots(k-m+1) a_k x^{k-m}$$

Thus $f \in C^{\infty}(I)$ and putting $x = 0$ in the last identity shows that $f^{(m)}(0) = m! a_m$ (since only the first term of the series makes a contribution if $x = 0$). Hence we have *determined the coefficients* in the power series expansion of f in terms of the values of the derivatives of f at 0. In recognition of its origins, $a_m = \frac{1}{m!} f^{(m)}(0)$ is called the mth *Maclaurin coefficient* of f, and $f(x) = \sum_{m=0}^{\infty} \frac{f^{(m)}(0)}{m!} x^m$ is the Maclaurin expansion of f. However, in the above there is nothing special about the role of 0, except that we found it convenient to express polynomials in terms of powers of x. If we expand C^{∞}-function f in terms of powers of $(x - a)$ instead, we obtain the general *Taylor series expansion* of f at a: $f(x) = \sum_{k=0}^{\infty} \frac{f^{(k)}(a)}{k!} (x-a)^k$, as long as the series converges. Here $\frac{f^{(k)}(a)}{k!}$ is the kth *Taylor coefficient* of f at a.

TUTORIAL PROBLEM 9.2

Naturally one may hope that the Taylor series of $f \in C^{\infty}(I)$ will converge to f at each point of I. That this need *not* be so is illustrated by a famous example due to Cauchy, namely

$$f(x) = \begin{cases} 0 & \text{if } x=0 \\ \exp(-\frac{1}{x^2}) & \text{if } x \neq 0 \end{cases}$$

Let us write $\exp x = e^x$ when it is convenient to simplify the notation. To find $f'(x)$ for $x \neq 0$ we use the Chain Rule and obtain

$$f'(x) = \left(\frac{2}{x^3}\right) e^{-\frac{1}{x^2}}$$

For $f'(0)$ we need to use the definition of the derivative directly:

$$f'_+(0) = \lim_{x \downarrow 0} \frac{f(x) - f(0)}{x - 0} = \lim_{x \downarrow 0} \frac{e^{-\frac{1}{x^2}}}{x} = \lim_{y \to \infty} y e^{-y^2}$$

and the last limit is 0 by Exercise on 4.4, since exponentials grow and decay faster than all powers. A similar argument shows that $f'_-(0) = 0$ also. Hence

f is differentiable everywhere and f' is given by

$$f'(x) = \begin{cases} 0 & \text{if } x=0 \\ \left(\frac{2}{x^3}\right)\exp\left(-\frac{1}{x^2}\right) & \text{if } x \neq 0 \end{cases}$$

The argument can now be repeated indefinitely to show that all the derivatives of f exist and are 0 at $x = 0$. The simplest method (though still fairly intricate) is to show by induction that

$$f^{(n)}(x) = \begin{cases} 0 & \text{if } x=0 \\ P_n\left(\frac{1}{x}\right)\exp\left(-\frac{1}{x^2}\right) & \text{if } x \neq 0 \end{cases}$$

where P_n is a polynomial of degree $3n$. The same limit argument as above then shows that $f^{(n)}(0) = 0$ for all n.

This means that all the coefficients of the Maclaurin expansion of f are 0, and hence that the expansion is identically 0. But then the expansion cannot represent the function f at any point other than 0, since $\exp(-\frac{1}{x^2}) \neq 0$.

It was with this example that Cauchy put paid to Lagrange's dream of avoiding infinitesimals by defining derivatives via series expansions, since it shows conclusively that even quite simply constructed functions such as f can fail to be represented by their series expansions. For us, the example means that we cannot expect power series to 'do the work' for us in all cases, and our search therefore turns to finding simple conditions on the functions which will exclude such 'monsters' as the above f.

Put differently: we are looking for sufficient conditions to ensure that the Taylor series of $f \in C^\infty$ will converge to $f(x)$. In doing so we shall also find forms of the *error* made in approximating f by a partial sum of its Taylor series, i.e. by a polynomial of given degree. This gives a measure of the *rate of convergence* of the series to $f(x)$. The error term $R_{N+1}(x) = f(x) - T_N^f(x)$ is called the *remainder of degree N*, and $T_N^f(x) = \sum_{n=0}^{N} \frac{f^{(n)}(a)}{n!}(x - a)^n$ is the nth degree *Taylor polynomial* of f expanded at a. Remarkably, an application of the *MVT* also gives an expression for R_{N+1} in terms of the values of a derivative of f:

● *Theorem 4—Taylor's Theorem*

Let I be an interval and suppose $f \in C^{N+1}(I)$. If $a, x \in I$ then there exists c between a and x such that

$$R_{N+1}(x) = \frac{(x - a)^{N+1}}{(N + 1)!} f^{(N+1)}(c)$$

PROOF
If $N = 0$ this is just the *MVT* again! Suppose that $a < x$ and set $M = \frac{(N+1)!}{(x-a)^{N+1}}\{f(x) - T_N^f(x)\}$. We need to show that $M = f^{(N+1)}(c)$ for some $c \in (a, x)$. We do this by applying Rolle's Theorem to the function g defined on $[a, x]$ by:

$$g(t) = -f(x) + f(t) + \sum_{k=1}^{N} \frac{(x - t)^k}{k!} f^{(k)}(t) + \frac{(x - t)^{N+1}}{(N + 1)!} M$$

This function is continuous in t on $[a, x]$, differentiable on (a, x), and $g(a) = g(x) = 0$ by our choice of M. By Rolle's Theorem there is $c \in (a, x)$ such that $g'(c) = 0$. Differentiating g w.r.t. t gives *(and you should check this carefully!)* $0 = g'(c) = \frac{(x-c)^N}{N!}(M - f^{(N+1)}(c))$, which proves the result.

● Proposition 4

Let $f \in C^\infty(I)$ and let R_{N+1} be as above. If the derivatives of f are uniformly bounded on I, that is, if there exists K such that $|f^{(n)}(x)| < K$ for all $x \in I$ and $n \in \mathbb{N}$, then $R_{N+1}(x) \to 0$ on I as $N \to \infty$.

PROOF

Again take $a < x$ in I and for each N write $R_{N+1}(x) = \frac{(x-a)^{N+1}}{(N+1)!} f^{(N+1)}(c_N)$ as in Taylor's Theorem. Then $c_N \in (a, x)$, and $|R_{N+1}(x)| \leq K \frac{(x-a)^{N+1}}{(N+1)!}$ by our hypothesis. This has the form $K \frac{A^m}{m!}$ for a fixed A, and $\frac{A^m}{m!} \to 0$ as $m \to \infty$ (e.g. since e^A is a convergent series!). The result follows, and under these conditions f is *represented* on I by its Taylor series expansion about a.

◉ Example 9

We expand $f(x) = \tan^{-1} x$ at $a = 1$ in order to estimate $\tan^{-1} 1.1$, using the fact that $\tan^{-1} 1 = \frac{\pi}{4}$. Differentiating f yields: $f'(x) = \frac{1}{1+x^2}$, $f''(x) = \frac{-2x}{(1+x^2)^2}$ and $f'''(x) = \frac{6x^2-2}{(1+x^2)^3}$. Thus the second degree Taylor polynomial of f expanded at 1 is given for $x > 1$ by:

$$T_2(x) = \frac{\pi}{4} + \frac{1}{10}\left(\frac{1}{2}\right) + \frac{1}{2!}\left(\frac{1}{10}\right)^2\left(\frac{-2}{4}\right)$$

and the remainder $f(x) - T_2(x) = R_3(x) = \frac{1}{3!}\left(\frac{1}{10}\right)^3\left\{\frac{6c^2-2}{(1+c^2)^3}\right\}$ for some c with $1 < c < x$. This error term can be estimated by noting that for $x = 1.1$, $\frac{6c^2-2}{(1+c^2)^3}$ lies between $\frac{6-2}{(2.1)^3}$ and $\frac{6(1.21)-2}{2^3}$, hence between $\frac{1}{2}$ and $\frac{2}{3}$, so that $R_3(1.1)$ lies between $\frac{1}{12\,000}$ and $\frac{2}{18\,000}$, thus providing a workable estimate of the accuracy of the approximation of $\tan^{-1} 1.1$ by $T_2(1.1) = \frac{\pi}{4} + \frac{1}{20} - \frac{1}{400}$.

TUTORIAL PROBLEM 9.3

Write a computer program to display the successive approximations to a given C^∞-function f by its Taylor polynomials $(T_n^f)_{n\geq 1}$ for $n = 1, 2, 3, \ldots$ You should observe how the intervals on which a good approximation is achieved gradually increase in length as n increases. (A Pascal program which illustrates this for some common functions is outlined in the Appendix.)

The 'usual suspects' exp, sin, cos, etc. furnish standard examples of the use of the Proposition, and you can now use Proposition 4 to prove easily that they are represented by their Maclaurin expansions throughout $\mathbb{R}$.

Example 10

The case of $\log(1+x)$ is slightly more delicate: we have seen that the (Maclaurin) series for $\log(1+x)$ is $x - \frac{x^2}{2} + \frac{x^3}{3} - \frac{x^4}{4} + \ldots = \sum_{n=1}^{\infty}(-1)^{n-1}\frac{x^n}{n}$ when $|x| < 1$. We can now check that this also holds when $x = 1$, which shows that $\log 2 = \sum_{n=1}^{\infty}(-1)^{n-1}\frac{1}{n}$, a result already derived, with rather more work, in Section 3.3. (The series does *not* make sense when $x = -1$; why not?)

To prove our claim, consider the remainders of the polynomial approximations to $f(x) = \log(1+x)$. These are $R_{N+1}(x) = \frac{x^{N+1}}{(N+1)!}f^{(N+1)}(c_N)$, and $c_N \in (0, x)$. Thus $R_{N+1}(x) = \frac{x^{N+1}}{(N+1)!}\frac{(-1)^N N!}{(1+c_N)^{N+1}}$, hence $R_{N+1}(1) = \frac{1}{(N+1)(1+c_N)^{N+1}} < \frac{1}{N+1}$, and thus $R_{N+1}(1) \to 0$ as $N \to \infty$. Therefore $\log 2 = f(1) = \sum_{n=1}^{\infty}(-1)^{n-1}\frac{1}{n}$.

In Taylor's Theorem the remainder can be given a variety of other forms which all have their uses in different applications. Further examples of the use of the theorem will be given in the Further Exercises at the end of the chapter.

EXERCISES ON 9.5

1. Use Taylor's Theorem to the following refinement of the Second Derivative Test for extreme points:
   If f is infinitely differentiable in the interval $N(a, \delta)$ for some $\delta > 0$, and if its first $(m-1)$ derivatives are all 0 at a, while $f^{(m)}(a) \neq 0$, then the following conclusions can be drawn:
   (i) if m is even and $f^{(m)}(a) > 0$, then f has a local minimum at a.
   (ii) if m is even and $f^{(m)}(a) < 0$, then f has a local maximum at a.
   (iii) if m is odd then a is not an extreme point for f.
   (*Hint*: write down the Taylor polynomial of degree $(m-1)$ for f at a, and note that $f(x) - f(a) = \frac{1}{m!}f^{(m)}(c)(x-a)^m$ for some c between x and a.)
   Apply this result to classify the extreme points of $f(x) = x^6 - x^4$.

2. Compute the ninth degree Taylor polynomial T_9 for $f(x) = \log(\frac{1+x}{1-x})$ expanded at 0. Use the remainder term to estimate $\log 3$.

3. Compute the third degree Taylor polynomial T_3 for $f(x) = \log\sec(x)$ expanded at 0. Use the remainder to show that the inequalities

$$x^4 \leq 12\log\sec x - 6x^2 \leq 8x^4$$

hold for $x \in (-\frac{\pi}{4}, \frac{\pi}{4})$.

Summary

In this chapter we have explored the consequences of the *completeness* of $\mathbb{R}$ for the concept of differentiability of real functions. Using the Boundedness Theorem we deduced the Mean Value Theorem, which allows us to deduce *global* properties of differentiable functions from a knowledge of the sign of the derivative. This provides a much stronger tool for finding extreme points and for determining the shape of the function. By repeating the arguments we were able to deduce Taylor's Theorem, which describes precisely how an infinitely differentiable function can be approximated by polynomials, whose coefficients are determined by the values of the derivatives of the function at a single point. These representations underpin the 'well-behaved' nature of functions defined by power series, which can be differentiated term by term throughout their domain. This allows us to complete our verification of the basic properties of the exponential, trigonometric and related functions.

Mean Value Theorems have many important applications to the approximation of functions, the determination of limits (e.g. in l'Hôpital's Rules) and the solution of equations – in the Further Exercises below we shall illustrate how they also enable us to refine the iterative techniques developed in earlier chapters.

FURTHER EXERCISES

1. Suppose the real function E satisfies the conditions:

   $$E(0) = 1, \qquad E'(x) = E(x), \qquad (x \in \mathbb{R}).$$

   By differentiating $g : x \mapsto E(x).E(a - x)$, show that the map g is constant, and find its value. Deduce that E satisfies:

   $$E(x + y) = E(x).E(y)$$

   at all points of $\mathbb{R}$. Write $e = E(1)$, and show that $e > 2$.
   Prove that $E(n) = e^n$, that E is strictly positive and increasing on $\mathbb{R}$ and hence find its range. Deduce that E coincides with the exponential function. (Compare with Example 7 in Chapter 6.)

2. Suppose $I = [a, b]$ and $f : I \mapsto I$ is differentiable. Use the MVT to show that:
   (i) the condition $|f'(x)| \le r < 1$ for all $x \in I$ is sufficient to ensure that f has a unique fixed point in I.
   (ii) there exists at least one point $c \in I$ such that $|f'(c)| \le 1$.
   (iii) if $f(x) = e^{-x}$ and $x_1 = 1$, $x_{n+1} = f(x_n)$ for $n \ge 1$, then (x_n) converges to a fixed point of f. Explain how this enables us to solve the equation $x + \log x = 0$. How do we know that there is exactly one solution?

3. Use Taylor's Theorem to prove the following:
   (i) for all $x \in \mathbb{R}$, $1 - \frac{x^2}{2} \le \cos x \le 1 - \frac{x^2}{2} + \frac{x^4}{24}$.
   (ii) for any $N \ge 1$, $e = \sum_{n=0}^{N} \frac{1}{n!} + \frac{e^{c_N}}{N!}$. Deduce that e must be irrational.

4. Find the following limits: $\lim_{x \to 0}(\cosh x)^{\frac{1}{x^2}}$, $\lim_{x \to \frac{\pi}{2}}(\sin x)^{\frac{1}{\sin x}}$.

5. *Binomial Theorem for general powers:*
   Show that the power series $\sum_{k \ge 0} \binom{\alpha}{k} x^k$ has radius of convergence 1 for all $\alpha \in \mathbb{R}$.
   Let $f(x) = \sum_{k=0}^{\infty} \binom{\alpha}{k} x^k$. Show that $(1 + x)f'(x) = \alpha f(x)$ for all x. By differentiating the ratio $\frac{f(x)}{(1+x)^\alpha}$, show that $f(x) = (1 + x)^\alpha$.

10 • The Riemann Integral

Although the Ancient Greeks were not very successful in dealing with tangents and instantaneous velocities, they were rather more adept at calculating areas and volumes. In this they consistently applied a form of approximation: trying to 'exhaust' the unknown shape by fitting it between shapes of known area or volume. One of their most lasting and impressive achievements in this direction was the collection of 'integration techniques' which were developed by Archimedes (287–212 BC) in order to calculate the areas and volumes of curvilinear figures. The ideas he developed are sometimes regarded as the first applications of 'calculus techniques' and the underlying principles are still relevant today. We recall a particularly striking example:

Example I—First turn of the spiral.

The *Archimedean spiral* is the locus of a point which starts at O on a line segment OA and moves uniformly along OA while the segment itself rotates uniformly about O. In polar coordinates, therefore, a typical point P on the spiral is given by the equation $r = a\theta$, where $r = |OP|$ is the radial distance and θ is the angle through which OA has rotated.

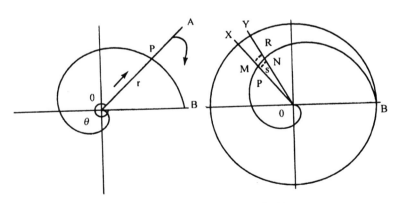

Fig. 10.1 Archimedes spiral

Archimedes calculates the area of the first full turn of the spiral: more precisely, he shows that this area is exactly $\frac{1}{3}$ of the area of the circle with radius OB, where B is the point on the spiral reached when $\theta = 2\pi$.

To see this, the circle OB is imagined as divided into n equal sectors, each suspending the angle $s = \frac{2\pi}{n}$ at O. The dividing lines defining these sectors cut the spiral in radii $r_k = a(ks)$, (i.e. when $\theta = ks$) for $k = 1, 2, \ldots, n$. Draw in the inscribed and circumscribed sectors to the spiral at each r_k. If P and R are the points where the spiral meets the kth sector OXY, then $OP = r_{k-1}$ and $OR = r_k$.

Denote the area of the kth sector OXY by s_k, that of the circumscribed sector OMR to the spiral by c_k and that of the inscribed sector OPN by i_k. Clearly $i_1 = 0$. The area $OPRB$ of the first full turn of the spiral lies between the areas $C = c_1 + c_2 + \ldots + c_n$ (the sum of the circumscribed areas) and $I = i_1 + i_2 + \ldots + i_n$ (that of the inscribed areas). But since $c_k = i_{k+1}$ for all $k < n$, it follows that $C - I = c_n = s_n = \frac{4\pi^3 a}{n}$, and we can make this difference as small as we please by taking n large enough. Now let S be the area of the circle OB. To find the ratio $\frac{C}{S}$, for example, note first that $\frac{c_k}{s_k} = \frac{r_k^2}{r_n^2}$. Hence

$$\frac{C}{S} = \frac{c_1 + c_2 + \ldots + c_n}{s_1 + s_2 + \ldots + s_n} = \frac{r_1^2 + r_2^2 + \ldots + r_n^2}{r_n^2 + r_n^2 + \ldots + r_n^2} = \frac{1^2 + 2^2 + \ldots + n^2}{n^2 + n^2 + \ldots + n^2}$$

The numerator is $\frac{1}{6}n(n+1)(2n+1)$, so that $\frac{C}{S} \to \frac{1}{3}$ (and similarly $\frac{I}{S} \to \frac{1}{3}$) as $n \to \infty$. But the limit of the areas C or I is just the area enclosed by the first turn of the spiral.

Ironically, Archimedes' description of the *method of discovery* which he employed to arrive at these results was lost during the Dark Ages: we only know of its existence because during a visit to Constantinople in 1906 the mathematical historian Heiberg discovered a tenth century copy of the manuscript – covered entirely under later (much more mundane) writings which had been superimposed on the same parchment at some time after the tenth century. Thus European mathematicians, struggling to develop a workable *calculus* in the sixteenth and seventeenth centuries, were denied the insights which Archimedes used to arrive at his formulae: in his more formal writings Archimedes invariably used the 'method of exhaustion', in which he approximated the area or volume to be measured by rectilinear areas or volumes (or, as with the spiral, by known curved areas such as sectors of circles). While this early version of a 'limit process' was also used by the early seventeenth century Italian, French and British mathematicians, in effect they had to invent the theory of integration anew. Only after considerable effort, and much later, did that theory reach an important stage in its development with the work of Bernhard Riemann (1826–66), whose construction of the integral of a real function is the content of this chapter.

10.1 Primitives and the 'arbitrary constant'

Given its long history, we could also ask why the integral, as a description of the area under the graph of a function f, should have anything to do with derivatives? In the Calculus we learn how to find the 'indefinite integral' $\int f(x)\,dx$ of f as a *function F* with $F' = f$. Thus F should be a *primitive* of f in the terminology of Chapter 9. However, we saw that primitives are not uniquely defined by a given f, since if F has derivative f, so does the function $G = F + c$ for any ('arbitrary') constant c. Conversely, given any two primitives F and G for f, it follows that $F' - G' = f - f = 0$, so that the difference $F - G$ *is* constant. We turn this to our advantage, by identifying the 'definite integral' $\int_a^b f(x)\,dx$ as the *number* $F(b) - F(a)$, and interpreting this number as the area under the graph of f over the interval $[a, b]$.

To show why the two operations of differentiation and integration are *inverse* to each other, Leibniz argues as follows in the 1670s: draw the graph of f as a straight

Fig. 10.2   Georg Friedrich Bernhard Riemann (1826–66)

line, and measure off ordinates $y_1, y_2, , y_n$ at equal distances along the x-axis. Imagining the distances between the ordinates as 'infinitely small' *units* of distance, he summed the ordinates $\sum_{k=1}^{n} y_k$ to obtain the area of the triangle 'under the graph' of f.

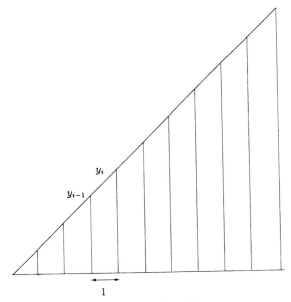

Fig. 10.3   Leibniz ordinates

Thus *summation* of ordinates provides us with the area. On the other hand, *subtraction* of successive ordinates, and dividing by the unit distance between them, gives $\frac{y_i - y_{i-1}}{1}$ as the slope of the graph of f near y_i, so that subtraction of ordinates leads to slopes of tangents, hence to values of the derivative of f. Since summation and subtraction are opposites, argues Leibniz, so must integration and differentiation be!

While this is hardly a convincing *proof*, it contains the germ of the idea of the *Fundamental Theorem of the Calculus*, which we shall use in the next chapter to justify and develop *techniques* for integration. First, however, we must turn to the underlying theory in order to be able to apply our results to a wide class of real functions.

10.2 Partitions and step functions: the Riemann integral

We shall define the integral for bounded real functions over bounded intervals. For the whole of this chapter we therefore fix a closed bounded interval $[a, b] \subset \mathbb{R}$ with $a < b$. A central role will be played by *step functions*, that is, functions which take only finitely many values, and assume each of their values on a subinterval of $[a, b]$. To make this more precise we need to introduce a little terminology first.

A *partition* of $[a, b]$ is a finite set $P = \{a_0, a_1, a_2, \ldots, a_n\}$ with

$$a = a_0 < a_1 < a_2 < \ldots < a_n = b$$

A partition Q is said to *refine* P if $P \subset Q$, i.e. if the finite set Q contains all the points of P. Given two partitions P_1 and P_2 of $[a, b]$, the partition $Q = P_1 \cup P_2$ therefore refines *both* P_1 and P_2.

A *step function* $\phi : [a, b] \mapsto \mathbb{R}$ is a function associated with a partition P of $[a, b]$ as follows: the range of ϕ is given by the finite set $\mathcal{R}_\phi = \{c_1, c_2, \ldots, c_n\}$ and for each $i \leq n$, ϕ takes the value c_i throughout the interval $(a_{i-1}, a_i]$. The values c_i and c_j need not be distinct for $i \neq j$; though in most examples it will be convenient to choose them so.

(In fact, the values of ϕ at the endpoints of each subinterval are not relevant to the theory, but we shall include them as shown in order to have ϕ properly defined throughout $[a, b]$.)

We shall use step functions to approximate the area under the graph of a general bounded real function f over the interval $[a, b]$. To this end it is useful to decide how step functions approximate certain classes of functions, such as continuous functions on $[a, b]$.

TUTORIAL PROBLEM 10.1

The collection of all step functions defined on $[a, b]$ is a *vector space*, that is, given two step functions ϕ and ψ, and any real numbers ('scalars') λ, μ, then the linear combination $\lambda\phi + \mu\psi$ is again a step function. Check this claim carefully, using refinements of partitions.

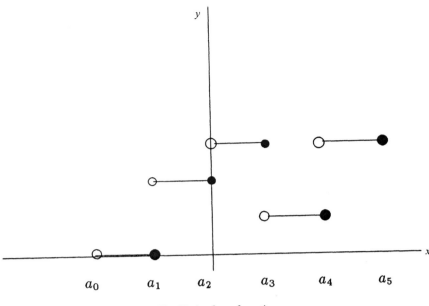

Fig.10.4 Step function

Continuous functions are now easy to handle, using the fact (see Theorem 4 in Chapter 7) that if f is continuous on the closed bounded interval $[a, b]$, then it is both bounded and *uniformly* continuous there:

● *Proposition 1*

Every continuous function $f : [a, b] \mapsto \mathbb{R}$ is a *uniform* limit of step functions, i.e. there exists a sequence of step functions $(\phi_n)_{n \geq 1}$ such that, given $\varepsilon > 0$, we can find $N \in \mathbb{N}$ such that, for all $x \in [a, b]$ and $n > N$, we have $|f(x) - \phi_n(x)| < \varepsilon$.

PROOF
Since f is uniformly continuous, given $\varepsilon > 0$ we can find $\delta > 0$ such that, for all $x, y \in [a, b]$, $|x - y| < \delta$ implies that $|f(x) - f(y)| < \varepsilon$. But $b - a$ is finite, so we can find $m \in \mathbb{N}$ such that $\frac{b-a}{m} < \delta$, and hence there is a partition $P = \{a_0, a_1, \ldots, a_m\}$ of $[a, b]$ such that each subinterval (a_{i-1}, a_i) has length less than δ. (We can take the partition points to be equidistant, i.e. set $a_i = a + \frac{i(b-a)}{m}$ for each $i \geq 1$, for example.) Then set $c_i = f(a_i)$ for $i = 1, 2, \ldots, m$. Then define a step function ϕ_ε by letting $\phi_\varepsilon(x) = c_i$ on $(a_{i-1}, a_i]$. By our choice of δ we have $|f(x) - \phi_\varepsilon(x)| < \varepsilon$ for all $x \in [a, b]$, since $|x - a_i| < \delta$ for all $x \in (a_{i-1}, a_i]$.

  This provides a step function ϕ_ε for each $\varepsilon > 0$. To construct a sequence as stated, repeat the procedure for $\varepsilon = 1, \frac{1}{2}, \frac{1}{3}, \ldots, \frac{1}{n},$. and write ϕ_n instead of $\phi_{\frac{1}{n}}$ for $n \geq 1$.

The Proposition shows that for each continuous function $f : [a, b] \mapsto \mathbb{R}$ and each $\varepsilon > 0$ we can find a *pair* of step functions $\phi = \phi_\varepsilon$ and $\psi = \psi_\varepsilon$ such that:

$$\phi < f < \psi \quad \text{and} \quad \psi(x) - \phi(x) = \varepsilon \quad \text{for all } x \in [a, b]$$

To see this, simply choose a step function θ which approximates f to within $\frac{\varepsilon}{2}$ and set $\phi = \theta - \frac{\varepsilon}{2}, \psi = \theta + \frac{\varepsilon}{2}$.

TUTORIAL PROBLEM 10.2

Suppose that (f_n) is a sequence of real functions defined on the interval $[a, b]$. A particular example of such a sequence was constructed as a sequence of step functions in Proposition 1. Discuss the difference between the statement of the Proposition (which we call *uniform convergence*) and the following statement, which asserts that (f_n) converges to a real function f *pointwise* on $[a, b]$: for each $x \in [a, b,]$ and each $\varepsilon > 0$ there exists $N \in \mathbb{N}$ such that $|f_n(x) - f(x)| < \varepsilon$ whenever $n > N$. Compare this with the difference between continuous and uniformly continuous functions on an interval $[a, b]$. (As before, it is instructive to pay close attention to the *order* in which the quantifiers are written!)

Can you construct a sequence of functions on $[0, 1]$ which converges to 0 pointwise but not uniformly?

To show that we can find the integral of a given bounded function f we will try to approximate f from above and below by step functions (much as Archimedes did, all those years ago!). Thus we first need to define what we mean by the integral of a step function, and then hope to construct the integral of f as some sort of limit of such integrals. The final step is not altogether simple, since we somehow need to take limits as the partitions involved become finer and finer. Fortunately, approximating simultaneously from above and below will make this easier to achieve. Our definitions are guided by this example and by what we have just shown to be true when f is continuous on $[a, b]$.

● *Definition 1*

Let $\phi : [a, b] \mapsto \mathbb{R}$ be a step function, with value c_i on the subinterval $(a_{i-1}, a_i]$ $(i = 1, 2, \ldots, n)$ of the partition P. The *integral of ϕ* is defined as

$$\int_a^b \phi = \sum_{i=1}^n c_i(a_i - a_{i-1})$$

The partition P gives rise to two particular step functions, ϕ_P^f and ψ_P^f, whose integrals define the *upper* and *lower Riemann sums* for f

$$\int_a^b \psi_P^f = \sum_{i=1}^n M_i \Delta a_i, \qquad \int_a^b \phi_P^f = \sum_{i=1}^n m_i \Delta a_i$$

where we write $\Delta a_i = a_i - a_{i-1}$ for brevity, and where, on the subinterval $(a_{i-1}, a_i]$ the step functions ψ_P^f and ϕ_P^f (respectively) take the values $M_i = \sup_{a_{i-1} \le x \le a_i} f(x)$ and $m_i = \inf_{a_{i-1} \le x \le a_i} f(x)$ for each $i \le n$. (Note that M_i and m_i are finite since f is assumed to be bounded!) Obviously we have $\int_a^b \phi_P^f \le \int_a^b \psi_P^f$ for every fixed partition P. We shall define the 'area under the graph' of f from these sums in such a way that for each partition P the upper sum is an overestimate and the lower sum an underestimate of the area.

◉ *Example 2*

Let us illustrate these definitions with a concrete example: set $f(x) = x^2$ and consider the function on the interval $[0, 1]$. A natural sequence of partitions is given by $P_n = \{0, \frac{1}{n}, \frac{2}{n}, \ldots, \frac{k}{n}, \ldots, 1\}$, so that, for fixed n, each subinterval $[a_{i-1}, a_i]$ has length $\Delta a_i = a_i - a_{i-1} = \frac{1}{n}$ for all $i \le n$. The corresponding step functions $\psi^f_{P_n}$ and $\phi^f_{P_n}$ are easily defined in this case: on the interval $[a_{i-1}, a_i]$ we have $m_i = f(\frac{i-1}{n}) = (\frac{i-1}{n})^2$ and $M_i = f(\frac{i}{n}) = (\frac{i}{n})^2$, since the continuous function f attains both its supremum and its infimum on each (closed and bounded) subinterval, and as f is increasing, these are attained at the endpoints. So for each $i = 1, 2 \ldots, n$ we set $\psi^f_{P_n}(x) = M_i$ and $\phi^f_{P_n}(x) = m_i$ for all $x \in (a_{i-1}, a_i]$.

Hence the upper and lower Riemann sums for the partition P_n are given simply by the sums:

$$\int_0^1 \psi^f_{P_n} = \sum_{i=1}^{n} \left(\frac{i}{n}\right)^2 \frac{1}{n} \quad \text{and} \quad \int_0^1 \phi^f_{P_n} = \sum_{i=1}^{n} \left(\frac{i-1}{n}\right)^2 \frac{1}{n}$$

To calculate these sums, recall that $\sum_{i=1}^{n} i^2 = \frac{1}{6} n(n+1)(2n+1)$. Hence the upper sum equals $\frac{1}{6n^3} n(n+1)(2n+1)$, while the lower sum is

$$\frac{1}{n^3 x} \left\{ \sum_{i=1}^{n} (i-1)^2 \right\} = \frac{1}{n^3} \sum_{j=1}^{n-1} j^2 = \frac{1}{6n^3} (n-1)(n) \left\{ 2(n-1) + 1 \right\}$$

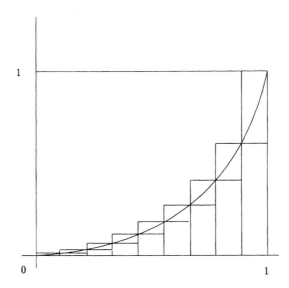

Fig. 10.5 Riemann sums

This holds for each fixed $n \ge 1$. Letting $n \to \infty$, that is, refining the partitions P_n successively by adding an extra point each time, we see that the sequences of upper and lower sums converge to the *same* limit:

$$\frac{1}{6n^3} n(n+1)(2n+1) = \frac{1}{6} \left(1 + \frac{1}{n}\right) \left(2 + \frac{1}{n}\right) \to \frac{1}{3}$$

while

$$\frac{1}{6n^3}(n-1)(n)\{2(n-1)+1\} = \frac{1}{6}\left(1-\frac{1}{n}\right)\left(2+\frac{1}{n}\right) \rightarrow \frac{1}{3}$$

The figure shows that, when such a sequence of partitions can be found, the common limit of their upper and lower sums should give us the area under the graph of f. We would like to turn this idea into the definition of the integral of f. However, we have not yet determined under what conditions the limit *exists*. Hence we shall adopt a slightly different approach as follows.

Fix a bounded function $f : [a, b] \mapsto \mathbb{R}$. We show that *refinements* of partitions always lead to *better approximations* to the area under the graph of f, that is, they produce smaller upper sums and larger lower sums (or more precisely – they don't make matters worse!):

To see this, let a partition P be given and suppose that Q refines P. Since both sets are *finite*, we can even assume that Q contains just one additional point, so that $P = \{a_0, a_1, \ldots, a_{k-1}, a_k, \ldots, a_n\}$ while $Q = \{a_0, a_1, \ldots, a_{k-1}, c, a_k, \ldots, a_n\}$. In other words, the subintervals of $[a, b]$ created by P and Q are identical except for the interval $[a_{k-1}, a_k]$, which is subdivided into the two intervals $[a_{k-1}, c]$ and $[c, a_k]$. Again, for any bounded $f : [a, b] \mapsto \mathbb{R}$ we define the 'upper' and 'lower' step functions ψ_P^f and ϕ_P^f relative to the partition P, as well as their counterparts relative to Q, i.e. ψ_Q^f and ϕ_Q^f. By definition, we obtain $\psi_P^f(x) = \sup_{x\in[a_{k-1},a_k]}f(x)$ for all $x \in (a_{k-1}, a_k]$, while $\psi_Q^f(x) = \sup_{x\in[a_{k-1},c]}f(x)$ on $(a_{k-1}, c]$ and $\psi_Q^f(x) = \sup_{x\in[c,a_k]}f(x)$ on $(c, a_k]$. Since both the latter suprema are taken over subsets of $[a_{k-1}, a_k]$ we must conclude that $\psi_Q^f(x) \le \psi_P^f(x)$ for all $x \in (a_{k-1}, a_k]$. Since the two functions coincide outside this interval we have proved that $\int_a^b \psi_Q^f \le \int_a^b \psi_P^f$ for these partitions P and Q. It is clear that a similar argument shows: $\int_a^b \phi_P^f \le \int_a^b \phi_Q^f$ also.

Thus we have proved our claim when Q has just one more partition point than P. Given a general refinement Q of P, we can apply this argument successively to the partitions produced by adding one new point from Q at a time to those of P, to obtain the same inequalities. So our claim holds in general.

We can go further: given *any* two partitions P, Q, consider the partition $R = P \cup Q$, which refines them both. It is now clear from the above arguments that:

$$m(b-a) \le \int_a^b \phi_P^f \le \int_a^b \phi_R^f \le \int_a^b \psi_R^f \le \int_a^b \psi_Q^f \le M(b-a)$$

where $m = \inf_{x\in[a,b]}f(x)$ and $M = \sup_{x\in[a,b]}f(x)$.

In other words: every lower sum is less than or equal to any upper sum, and (in $\mathbb{R}$) the set of all possible upper and lower sums is bounded above by $M(b-a)$ and below by $m(b-a)$. (This is one reason why we have restricted ourselves to bounded functions on bounded intervals.)

Denote the set of all upper sums for f by $\mathcal{U}(f)$ and the set of all lower sums by $\mathcal{L}(f)$. Both are thus bounded subsets of $\mathbb{R}$, so that, in particular, the supremum of $\mathcal{L}(f)$ and the infimum of $\mathcal{U}(f)$ exist as real numbers. This allows us to complete our general definition of the Riemann integral of f:

● Definition 2

The bounded function $f : [a, b] \mapsto \mathbb{R}$ is *Riemann-integrable over* $[a, b]$ if its *upper integral* $\overline{\int}_a^b f = \inf \mathcal{U}(f)$ coincides with its *lower integral* $\underline{\int}_a^b f = \sup \mathcal{L}(f)$. In that case we call their common value the *Riemann integral of f over* $[a, b]$ and denote this number by $\int_a^b f$ – or alternatively by $\int_a^b f(x) dx$.

Note that since any upper sum dominates all lower sums it must also dominate the lower integral. Since this holds for each upper sum it also holds for the upper integral, so that we have proved:

$$\underline{\int}_a^b f \le \overline{\int}_a^b f \quad \text{for each bounded function } f$$

Since the elements of $\mathcal{U}(f)$ are upper sums we can also write $\overline{\int}_a^b f = \inf\{\int_a^b \psi_P^f : P$ is a partition of $[a, b]\}$ and similarly $\underline{\int}_a^b f = \sup\{\int_a^b \phi_P^f : P$ is a partition of $[a, b]\}$. This illustrates the general nature of the definition: we are required to know the structure of upper and lower sums for *every* partition of $[a, b]$, which is asking a great deal, as the following example shows:

● Example 3

Consider $f = 1_{\mathbb{Q}}$ over the interval $[0, 1]$: since for every partition P of $[0, 1]$ every subinterval $[a_{i-1}, a_i]$ contains both rational and irrational numbers, we see at once that each ϕ_P^f is identically 0 (as f takes the value 0 at each irrational), while each ψ_P^f is identically 1 (as f takes the value 1 at each rational point). Hence every upper sum is 1 and every lower sum is 0, so that the upper integral $\overline{\int}_0^1 1_{\mathbb{Q}} = 1$, while the lower integral $\underline{\int}_0^1 1_{\mathbb{Q}} = 0$. Therefore f is *not* Riemann-integrable.

Although the number of possible partitions of an interval is huge, we saw in Example 2 that we could construct a *sequence* of partitions whose upper and lower sums converge to the same limit. This provides a useful technique for evaluating integrals directly, since the common limit of the sequences must also be the common value of $\inf \mathcal{U}(f)$ and $\sup \mathcal{L}(f)$: each element of $\mathcal{U}(f)$ (i.e. each upper sum) is at least as great as each element of $\mathcal{L}(f)$ (i.e. each lower sum) and the partitions P_n produce upper and lower sums that differ by as little as we please. We shall use this idea in the next section to produce more usable criteria for establishing when a given function is Riemann-integrable.

EXERCISES ON 10.2

1. Show that the function $x \mapsto x^3$ is Riemann-integrable over $[0, 1]$ and find the value of $\int_0^1 x^3 dx$ directly from the definition of the Riemann integral. (Recall that $\sum_{k=1}^n k^3 = \left(\frac{n(n+1)}{2}\right)^2$.)
2. Suppose that $f : [a, b] \mapsto \mathbb{R}$ is integrable, and let $h = -f$. Let P be a partition of $[a, b]$. Show that U_P is the upper sum for f relative to P if and only if $-U_P$ is the lower sum for h relative to P. Show similarly that if L_P is the lower sum for f

then $-L_P$ is the upper sum for h. Deduce that h is integrable and that $\int_a^b h = -\int_a^b f$.

10.3 Criteria for integrability

One of our objectives is to prove results which show that *every* function with a certain property is Riemann-integrable, so that we do not laboriously have to plough through summation arguments for each function separately. (Of course, we cannot yet expect that this procedure will tell us the *value* of the integral in each case – though we shall discuss techniques to achieve this in the next chapter.) For this purpose we now establish *criteria* which can easily be checked and which guarantee that a given function f is (Riemann-) integrable. We state the principal criterion in two ways: using partitions and using step functions.

● *Theorem 1—Riemann's criterion for integrability* ——

A bounded function $f : [a, b] \mapsto \mathbb{R}$ is Riemann-integrable if and only if for every $\varepsilon > 0$ there exists a partition P_ε whose upper and lower sums differ by less than ε.

In other words: f is Riemann-integrable if and only if there are step functions ψ and ϕ (namely $\psi = \psi^f_{P_\varepsilon}$ and $\phi = \phi^f_{P_\varepsilon}$) such that $\phi(x) \leq f(x) \leq \psi(x)$ for all $x \in [a, b]$ and $\int_a^b \psi - \int_a^b \phi < \varepsilon$.

PROOF

If, for each $\varepsilon > 0$ a partition P_ε can be found with the stated properties, then $\overline{\int_a^b} f \leq \int_a^b \psi^f_{P_\varepsilon} \leq \varepsilon + \int_a^b \phi^f_{P_\varepsilon} \leq \varepsilon + \underline{\int_a^b} f$ by definition of the upper and lower integral of f. But we have shown that $\underline{\int_a^b} f \leq \overline{\int_a^b} f$, hence $|\overline{\int_a^b} f - \underline{\int_a^b} f| < \varepsilon$ for all $\varepsilon > 0$. Since this holds for all $\varepsilon > 0$ it follows that the upper and lower integrals are equal, hence f is integrable.

Conversely, if f is integrable, and $\varepsilon > 0$ is given, we can find a partition P_1 whose upper sum approximates the upper integral to within $\frac{\varepsilon}{2}$, and a partition P_2 whose lower sum approximates the lower integral similarly. The partition $P_\varepsilon = P_1 \cup P_2$ refines both P_1 and P_2 hence its lower sum approximates the lower integral better than that of P_2 and its upper sum approximates the upper integral better than that of P_1. Since f is integrable, its upper and lower integrals both equal $\int_a^b f$, hence:

$$\int_a^b f - \frac{\varepsilon}{2} < \int_a^b \phi^f_{P_2} \leq \int_a^b \phi^f_{P_\varepsilon} \leq \int_a^b \psi^f_{P_\varepsilon} \leq \int_a^b \psi^f_{P_1} < \int_a^b f + \frac{\varepsilon}{2}$$

which shows that $\int_a^b \psi^f_{P_\varepsilon} - \int_a^b \phi^f_{P_\varepsilon} < \varepsilon$, as required.

● *Example 4*

We calculate $\int_0^1 f(x)\,dx$ when $f(x) = \sqrt{x}$: our immediate problem is that square roots are hard to find except for perfect squares. Therefore we shall take partition points which are perfect squares, even though this means that the interval lengths of the different intervals do not stay the same (there is nothing to say that they should do, of course, even if it often simplifies the calculations!). In fact, take the sequence of partitions

$$P_n = \left\{ 0, \left(\frac{1}{n}\right)^2, \left(\frac{2}{n}\right)^2, \ldots, \left(\frac{i}{n}\right)^2, \ldots, 1 \right\}$$

and consider the upper and lower sums, using the fact that f is increasing on each interval $[(\frac{i-1}{n})^2, (\frac{i}{n})^2]$. We obtain:

$$\int_0^1 \psi_{P_n}^f = \sum_{i=1}^n \frac{i}{n} \left\{ \left(\frac{i}{n}\right)^2 - \left(\frac{i-1}{n}\right)^2 \right\} = \frac{1}{n^3} \sum_{i=1}^n (2i^2 - 1)$$

$$\int_0^1 \phi_{P_n}^f = \sum_{i=1}^n \left(\frac{i-1}{n}\right) \left\{ \left(\frac{i}{n}\right)^2 - \left(\frac{i-1}{n}\right)^2 \right\} = \frac{1}{n^3} \sum_{i=1}^n (2i^2 - 3i + 1)$$

Hence

$$\int_0^1 \psi_{P_n}^f - \int_0^1 \phi_{P_n}^f = \frac{1}{n^3} \sum_{i=1}^n (2i - 1) = \frac{1}{n^3} \{n(n+1) - n\} = \frac{1}{n}$$

By choosing n large enough, we can make this difference less than any given $\varepsilon > 0$, hence f is integrable. The integral must be $\frac{2}{3}$, since the sequences of upper and lower sums both converge to this value, as is easily seen.

Let us return to the representation of the integral as an 'area operator', that is, a mapping $A : f \mapsto A(f)$ which associates to each bounded function $f : [a, b] \mapsto \mathbb{R}$ the area under the graph of f. Two obvious requirements for such a function are:

(i) if $m = \inf_{x \in [a,b]} f(x)$ and $M = \sup_{x \in [a,b]} f(x)$ then

$$m(b - a) \le A(f) \le M(b - a)$$

(ii) A is additive over intervals, i.e. if $c \in (a, b)$ is given and if $f_{|[a,c]}$ and $f_{|[c,b]}$ denote the *restrictions* of f to the subintervals $[a, c]$ and $[c, b]$ respectively, then

$$A(f) = A(f_{|[a,c]}) + A(f_{|[c,b]})$$

We have already seen that the mapping $A(f) = \int_a^b f$ satisfies the first of these claims. To prove that the second property also holds for the Riemann integral, we have:

● *Proposition 2*

The real function f is integrable over $[a, b]$ if and only if for each $c \in (a, b)$ its restrictions $f_{|[a,c]}$ and $f_{|[c,b]}$ are integrable (over $[a, c]$ and $[c, b]$ respectively). Moreover, the identity $\int_a^b f = \int_a^c f + \int_c^b f$ holds whenever either side exists.

PROOF
If f is integrable over $[a, b]$ then for $\varepsilon > 0$ the Riemann criterion provides a partition P_ε whose upper and lower sums differ by less than ε. We can assume without loss that $c \in P_\varepsilon$, since, if c is not a partition point, we can add it to P_ε, so that the new partition refines P_ε and hence yields even better approximations. So take $P_\varepsilon = \{a_0, a_1, \ldots, a_n\}$ with $a_m = c$. With $M_k = \sup_{x \in [a_{k-1}, a_k]} f(x)$ and

$m_k = \inf_{x \in [a_{k-1}, a_k]} f(x)$ for $k \leq n$, the upper sums for f over $[a, c]$ and $[c, b]$ become:

$$\int_a^c \psi^f_{P_\varepsilon} = \sum_{k=1}^m M_k \Delta a_k, \qquad \int_c^b \psi^f_{P_\varepsilon} = \sum_{k=m+1}^n M_k \Delta a_k$$

while the corresponding lower sums are

$$\int_a^c \phi^f_{P_\varepsilon} = \sum_{k=1}^m m_k \Delta a_k, \qquad \int_c^b \phi^f_{P_\varepsilon} = \sum_{k=m+1}^n m_k \Delta a_k$$

Since $M_k \geq m_k$ for all $k \geq 1$, and $\sum_{k=1}^n (M_k - m_k) \Delta a_k < \varepsilon$, it follows that $\sum_{k=1}^m (M_k - m_k) \Delta a_k < \varepsilon$ and $\sum_{k=m+1}^n (M_k - m_k) \Delta a_k < \varepsilon$. Hence the upper and lower sums differ by less than ε on each interval $[a, c]$ and $[c, b]$, so that the restrictions $f_{|[a,c]}$ and $f_{|[c,b]}$ are both integrable.

To prove the converse, we find partitions P_1 of $[a, c]$ and P_2 of $[c, b]$ such that the upper and lower sums differ by less than $\frac{\varepsilon}{2}$ on each interval. Setting $P_\varepsilon = P_1 \cup P_2$ provides a partition of $[a, b]$ of exactly the same form as before, i.e. $P_\varepsilon = \{a_0, a_1, \ldots, a_n\}$ with $a_m = c$. Now the upper sums are additive for this partition, i.e. $\int_a^b \psi^f_{P_\varepsilon} = \int_a^c \psi^f_{P_\varepsilon} + \int_c^b \psi^f_{P_\varepsilon}$, and similarly for the lower sums. Hence the upper and lower sums for f on $[a, b]$ differ by less than ε, so f is integrable on $[a, b]$.

Finally, by construction the upper and lower sums have just been shown to be interval-additive, so that if f is integrable and, for given $\varepsilon > 0$, $P = P_\varepsilon$ as above, then $\int_a^b f$ lies between $\int_a^b \phi^f_P$ and $\int_a^b \psi^f_P < \int_a^b \phi^f_P + \varepsilon$. But we also have:

$$\int_a^b \phi^f_P = \int_a^c \phi^f_P + \int_c^b \phi^f_P \leq \int_a^c f + \int_c^b f$$

$$\leq \int_a^c \psi^f_P + \int_c^b \psi^f_P < \int_a^c \phi^f_P + \int_c^b \phi^f_P + \varepsilon$$

Therefore $|\int_a^b f - (\int_a^c f + \int_c^b f)| < \varepsilon$. Since this holds for all $\varepsilon > 0$, it follows that the two quantities are equal.

The additivity formula remains true even if c does not lie between a and b. We ensure this by a definition: our integrals have so far been defined only if $a < b$. Now assume that $b < a$ and *define* $\int_a^b f = -\int_b^a f$, where the right-hand side makes sense as the Riemann integral of f over the interval $[b, a]$. The additivity formula then makes sense with a, b, c given in any order, as long as f is integrable over all the intervals concerned. For example, suppose that $b < a < c$ and f is integrable over $[b, c]$. Then f is also integrable over $[b, a]$ and $[a, c]$, by Proposition 2, and $\int_b^c f = \int_b^a f + \int_a^c f$. But this can be rewritten as follows: $\int_c^b f = -\int_b^c f$, and $\int_a^b f = -\int_b^a f$, hence $\int_a^b f = \int_a^c f + \int_c^b f$ as claimed.

Finally, note that if $c = a$, we obtain: $\int_a^b f = \int_a^a f + \int_a^b f$, which shows that $\int_a^a f = 0$.

Example 5

The function $f(x) = \begin{cases} x^2 & 0 \leq x \leq 1 \\ x & 1 \leq x \leq 2 \end{cases}$ is now easily seen to be integrable over $[0, 2]$, since the restrictions to $[0, 1]$ and $[1, 2]$ are integrable. Moreover, as we can easily check from the definitions:

$$\int_0^2 f = \int_0^1 x^2 \, dx + \int_1^2 x \, dx = \frac{1}{3} + \frac{3}{2} = \frac{11}{6}$$

EXERCISES ON 10.3

1. Show directly from the definition that the integral is monotone, i.e. if $f(x) \leq g(x)$ for all $x \in [a, b]$, then $\int_a^b f \leq \int_a^b g$.
2. Use the previous Exercise and the two 'area function' properties of the integral to prove that the map $F : [a, b] \mapsto \mathbb{R}$ given by $x \mapsto \int_a^x f$ is Lipschitz-continuous, with Lipschitz constant $M = \sup_{x \in [a,b]} |f(x)|$.
   (In other words, show that: $|F(x) - F(y)| \leq M|x - y|$.)
   (*Hint*: assume $y \leq x$ and set $h = x - y$. Show that $|F(x) - F(y)| = |\int_y^{y+h} f| \leq Mh$.)

10.4 Classes of integrable functions

We have done all the preparatory work to prove that two very large classes of bounded real functions are integrable over every bounded interval $[a, b]$:

● *Proposition 3*

Every bounded monotone function is integrable over $[a, b]$.

PROOF
Suppose that $f : [a, b] \mapsto \mathbb{R}$ is decreasing. Partition $[a, b]$ into n intervals of equal length, so that the partition $P = \{a_0, a_1, \ldots, a_n\}$ has $a_k = a + \frac{k(b-a)}{n}$ for $k \leq n$. The difference of the upper and lower sums for f relative to P is therefore

$$\int_a^b \psi_P^f - \int_a^b \phi_P^f = \frac{b-a}{n} \sum_{k=1}^n \{f(a_k) - f(a_{k-1})\} = \frac{b-a}{n} \{f(b) - f(a)\}$$

As the distances $(b - a)$ and $\{f(b) - f(a)\}$ are bounded, it is clear that for given $\varepsilon > 0$ we can choose n so large that $\frac{b-a}{n}\{f(b) - f(a)\} < \varepsilon$. Hence f satisfies Riemann's criterion for integrability.

If f is increasing on $[a, b]$ then $-f$ is decreasing, so the integrability of f will follow from the above and from Exercise on 10.2(2).

● *Theorem 2* ─────────────────────────────

Every continuous function $f : [a, b] \mapsto \mathbb{R}$ is integrable over $[a, b]$.

PROOF
This is an immediate consequence of the remark following Proposition 1. Given $\varepsilon > 0$ let $\eta = \frac{\varepsilon}{2(b-a)}$ and find step functions $\phi = \phi_\varepsilon$ and $\psi = \psi_\varepsilon$ such that $\phi \leq f \leq \psi$ and $\psi - \phi = \eta$. Thus, by construction of the integral for step functions, and since

for the constant η we have $\int_a^b \eta = \eta(b-a)$,

$$\int_a^b \psi - \int_a^b \phi = \int_a^b (\psi - \phi) = \eta(b-a) < \varepsilon$$

Hence, if $P_\varepsilon = P_1 \cup P_2$, where P_1 and P_2 are the partitions generated by ψ and ϕ, then the upper and lower sums of P_ε differ by less than ε. So f is integrable.

We can extend the Proposition to include functions which have only finitely many points of discontinuity: suppose that $P = \{a_0, a_1, \ldots, a_n\}$ is a partition of $[a, b]$ and that $f : [a, b] \mapsto \mathbb{R}$ is continuous *except* possibly at points of P. In other words, on each subinterval (a_{i-1}, a_i) we have a continuous function f_i and $f_{|(a_{i-1},a_i]} = f_i$ for each $i \le n$. We call such an f a *piecewise continuous* function on $[a, b]$. Since the integral is additive over intervals we can then define $\int_a^b f = \sum_{i=1}^n \int_{a_{i-1}}^{a_i} f_i$, and this is easily seen to be consistent with our earlier definitions.

For example, the bounded piecewise continuous function $f(x) = x - [x]$ has

$$\int_0^4 f = \int_0^1 x \, dx + \int_1^2 (x-1)dx + \int_2^3 (x-2)dx + \int_3^4 (x-3)dx = 2$$

We can extend this train of thought much further:

TUTORIAL PROBLEM 10.3

Explore the ideas below to discuss how we may extend Proposition 2 to apply to bounded functions which have discontinuities:

(i) suppose that for each $\varepsilon > 0$, the set A of points where f fails to be continuous can be contained in a *finite* sequence of intervals of total length less than ε; that is $A \subset \cup_{k=1}^m I_k$ where the intervals $I_k = [a_k, b_k]$ satisfy: $\sum_{k=1}^m (b_k - a_k) < \varepsilon$. Show that f is integrable on $[a, b]$. (*Hint:* consider how we tackled Example 5.)

(ii) There is life beyond this book: a taster for the dedicated!

Consult the literature to discover how things change when we replace the finite sequence of intervals in (i) by an *infinite* sequence of intervals (I_k) of total length less than ε: we define a set $A \subset \mathbb{R}$ to be a *null set* if for every $\varepsilon > 0$ we can construct a sequence (I_k) of intervals $I_k = [a_k, b_k]$ such that $\sum_{k=1}^\infty (b_k - a_k) < \varepsilon$. Note in particular that this extension of the above ideas provides a *characterization* of the set of Riemann-integrable functions, namely:

Lebesgue's criterion for Riemann integrability A bounded function $f : [a, b] \mapsto \mathbb{R}$ is Riemann-integrable if and only if f is continuous except possibly on some null set.

A proof of the last result is beyond the scope of this book. You may find a proof in Section 7.26 of *Mathematical Analysis* by Tom Apostol (2nd edition). One can then see that the Dirichlet function introduced as Example 7 in Chapter 6 satisfies this criterion, but not the condition discussed in (i) above: this function is discontinuous at each rational point, and continuous at each irrational. It turns out that the set $\mathbb{Q}$ is a null set, but it cannot be contained within a *finite* sequence of intervals of arbitrarily small total length. These subtleties were the cause of much fundamental research in the late nineteenth century.

EXERCISES ON 10.4

1. Explain carefully why in each of the following cases f is integrable over $[a, b]$:
   (i) $f(x) = |x^3|$, $[a, b] = [-1, 1]$
   (ii) $f(x) = sgn(x)$, $[a, b] = [-2, 2]$
   (iii) $f(x) = [x^3]$, $[a, b] = [0, 2]$.
   Suggest a value for the integral in each case.
2. Calculate $\int_0^3 f$ when $f(x) = 1$ for $0 \le x < 1$, $f(x) = (x - 1)^2$ for $1 \le x < 2$ and $f(x) = \sqrt{(x - 2)}$ for $2 \le x \le 3$.

10.5 Properties of the integral

We now derive some simple algebraic and order properties of the integral. We deal with the latter first. In particular, *you* should already have shown in Exercise on 10.3(1) that the integral is *monotone* – but here is a quick proof just to make sure: if $f \le g$ on $[a, b]$ and both functions are integrable, then $\int_a^b f = \underline{\int_a^b} f = \sup\{\int_a^b \phi_P^f : P$ is a partition of $[a, b]\}$. But since $f \le g$, the (lower) step functions ϕ_P^f and ϕ_P^g associated with P must satisfy $\phi_P^f \le \phi_P^g$, and hence the same inequality holds for their integrals, and thus also for the lower integrals of f and g. Therefore $\underline{\int_a^b} f \le \underline{\int_a^b} g$.

This leads to a 'Triangle Inequality' for integrals: if f and $|f|$ are integrable, then

$$\left| \int_a^b f \right| \le \int_a^b |f|$$

The proof is simple: we know that $-|f| \le f \le |f|$, so that since the integral is monotone, $-\int_a^b |f| \le \int_a^b f \le \int_a^b |f|$, and since $|f| \ge 0$ implies $\int_a^b |f| \ge 0$, the result follows.

To see that the integral is *linear* we exploit the obvious linearity of the integrals defined for step functions similarly. Here a simple consequence of Riemann's criterion is useful when expressed in terms of step functions: if f is integrable, we can find sequences of step functions (ϕ_n) and (ψ_n) with $\phi_n \le f \le \psi_n$ and $\int_a^b (\psi_n - \phi_n) \to n$ as $n \to \infty$. (Simply let $\varepsilon = \frac{1}{n}$ successively in the Riemann criterion.) This leads to the following:

● *Proposition 4*

Let f and g be integrable over $[a, b]$ and let $\alpha, \beta \in \mathbb{R}$ be given. Then the linear combination $\alpha f + \beta g$ is integrable over $[a, b]$ and $\int_a^b (\alpha f + \beta g) = \alpha \int_a^b f + \beta \int_a^b g$.

PROOF
We choose sequences of step functions as outlined above: $\phi_n^f \le f \le \psi_n^f$ and $\int_a^b (\psi_n^f - \phi_n^f) \to 0$ as $n \to \infty$, and similarly $\phi_n^g \le g \le \psi_n^g$ and $\int_a^b (\psi_n^g - \phi_n^g) \to 0$ as

$n \to \infty$. By the sandwich principle this means, in particular, that $\int_a^b \phi_n^f \to \int_a^b f$ and $\int_a^b \phi_n^g \to \int_a^b g$ as $n \to \infty$. Here we have used the fact that the integrals are linear and monotone. But now $\phi_n^f + \phi_n^g \le f + g \le \psi_n^f + \psi_n^g$ and

$$\int_a^b (\psi_n^f + \psi_n^g) - (\phi_n^f + \phi_n^g) = \int_a^b (\psi_n^f - \phi_n^f) + \int_a^b (\psi_n^g - \phi_n^g) \to 0$$

as $n \to \infty$. Hence $f + g$ is integrable and its integral equals

$$\lim_{n\to\infty} \int_a^b (\phi_n^f + \phi_n^g) = \lim_{n\to\infty} \int_a^b \phi_n^f + \lim_{n\to\infty} \int_a^b \phi_n^g = \int_a^b f + \int_a^b g$$

The proof that αf is integrable and $\int_a^b \alpha f = \alpha \int_a^b f$ is similar to the above when $\alpha \ge 0$. For $\alpha < 0$ note that $\alpha f = (-\alpha)(-f)$ and recall that $\int_a^b (-f) = -\int_a^b f$, so that (since $-\alpha > 0$) we have, finally:

$$\int_a^b \alpha f = \int_a^b (-\alpha)(-f) = (-\alpha) \int_a^b (-f) = \alpha \int_a^b f$$

This completes the proof of the Proposition.

We can now improve our statement of the 'Triangle Inequality' for integrals: for the proof given above we needed to know in advance that f and $|f|$ are *both* integrable. But we can now show that if f is integrable, then so is $|f|$. First we need some useful new functions which are easy to analyse.

TUTORIAL PROBLEM 10.4

Show that the non-negative functions $f^+ = \max(f, 0)$ and $f^- = \max(-f, 0)$ (which we call the *positive* and *negative part of* f, respectively) satisfy the equations:

$$f = f^+ - f^- \quad \text{and} \quad |f| = f^+ + f^-$$

Show further that for any real functions f, g we have:

$$\max(f, g) = \tfrac{1}{2}\{|f + g| + |f - g|\}, \quad \min(f, g) = \tfrac{1}{2}\{|f + g| - |f - g|\}.$$

Illustrate your proofs with sketches for $f(x) = \sin x$ and $g(x) = \cos x$ on $[0, 2\pi]$.

Let us now show that if f is integrable, then so is f^+: again, for $\varepsilon > 0$ find step functions ϕ, ψ satisfying $\phi \le f \le \psi$ and $\int_a^b (\psi - \phi) < \varepsilon$. But taking the positive parts of these step functions, we have $\phi^+ \le f^+ \le \psi^+$ and moreover $\psi^+ - \phi^+ \le \psi - \phi$ (from the definition of the positive part!), hence also $\int_a^b (\psi^+ - \phi^+) \le \int_a^b (\psi - \phi) < \varepsilon$. This shows that $f^+ = \max(f, 0)$ is integrable when f is. It is immediately obvious that $f^- = \max(-f, 0)$ is also integrable, since $-f$ is integrable, so the above argument applies. Finally we conclude from Proposition 4 that $|f|$ is also integrable.

The above properties also provide a limit theorem for Riemann integrals:

• *Theorem 3*

Suppose that (f_n) is a sequence of (Riemann-)integrable functions and that $f_n \to f$ *uniformly* on $[a, b]$ as $n \to \infty$ (in the sense described in Proposition 1 and Tutorial Problem 2, namely, given $\varepsilon > 0$ we can find $N \in \mathbb{N}$ such that for all $n > N$ and all $x \in [a, b]$, $|f_n(x) - f(x)| < \varepsilon$). Then f is also integrable and $\lim_{n\to\infty} \int_a^b f_n = \int_a^b f$.

PROOF
Given $\varepsilon > 0$, choose N so that $n > N$ implies $|f_n(x) - f(x)| < \varepsilon$ for all $x \in [a, b]$. Fix such an n and choose step functions ϕ, ψ with $\phi \leq f_n \leq \psi$, $\int_a^b (\psi - \phi) < \varepsilon$. Since $f_n - \varepsilon < f < f_n + \varepsilon$, we have $\phi - \varepsilon < f < \psi + \varepsilon$. Now $\phi - \varepsilon$ and $\psi + \varepsilon$ are step functions and

$$\int_a^b \{(\psi + \varepsilon) - (\phi - \varepsilon)\} = \int_a^b (\psi - \phi) + \int_a^b 2\varepsilon < \varepsilon\{2(b - a) + 1\}$$

This can again be made arbitrarily small, hence f satisfies Riemann's criterion. Now we can estimate:

$$\left| \int_a^b f_n - \int_a^b f \right| = \left| \int_a^b (f_n - f) \right| \leq \int_a^b |f_n - f| < \int_a^b \varepsilon = \varepsilon(b - a)$$

This completes the proof, since ε was arbitrary.

TUTORIAL PROBLEM 10.5

(Refer to Tutorial Problem 10.2 once more for the definitions you need below.) Show that $f_n(x) = ne^{-nx}$ converges *pointwise* to 0 on the interval $(0, 1)$, but that the sequence of integrals $(\int_0^1 f_n)$ converges to 1. Why does this not conflict with Theorem 3?

Finally, we prove a very simple Mean Value Theorem for integrals which is often useful when we need to estimate integrals whose exact value we cannot find easily:

• *Proposition 5*

Let f be continuous and let g be integrable and non-negative on $[a, b]$. Then there exists $c \in [a, b]$ such that

$$\int_a^b f \cdot g = f(c) \int_a^b g$$

In particular (with $g(x) = 1$ on $[a, b]$): $\int_a^b f = f(c)(b - a)$.

PROOF
By the Boundedness Theorem in Chapter 7, the continuous function f attains both its maximum M and its minimum m at points of $[a, b]$, so that

$mg(x) \le f(x).g(x) \le Mg(x)$ for all $x \in [a, b]$. As the integral is linear and monotone, this implies that

$$m \int_a^b g \le \int_a^b f.g \le M \int_a^b g$$

We can assume that $\int_a^b g > 0$, since otherwise $\int_a^b f.g = 0$ would follow from the above inequalities and the result follows trivially.

Write $K = \frac{\int_a^b f.g}{\int_a^b g}$. The above inequalities show that $m \le K \le M$, so that the *IVT*, applied to f on $[a, b]$, yields a point $c \in [a, b]$ such that $f(c) = K$. This proves our claim.

It is often useful to reformulate this result; let $h = b - a$, so that $c = a + \theta h$, where $0 \le \theta \le 1$. Then the Proposition claims that:

$$\int_a^{a+h} f.g = f(a + \theta h) \int_a^{a+h} g$$

In this form, and with the convention $\int_a^b f = -\int_b^a f$ when $b < a$ (see the remarks following Proposition 2) the result remains true for $h < 0$ also.

EXERCISES ON 10.5

1. Calculate $\int_0^1 (3x^3 - 2x^2 + 5\sqrt{x} - 1)dx$.
2. Show that for each $n \ge 1$, $f_n : x \mapsto \frac{\cos nx}{1+x^2}$ is integrable over $[0, 1]$. Explain why $I_n \to 0$, where $I_n = \int_0^1 f_n$.

Summary

We have introduced the Riemann integral of a bounded function f as an area function, describing the area under the graph of f. The construction proceeded via step functions, for which the area is easily calculated as a finite sum of 'areas of rectangles'. For more general f (in particular, those functions which can be *uniformly* approximated by step functions) it is then possible to define $\int_a^b f$ as a limit of the integrals of step functions. This must be done with some care, and it was simplest to construct approximating step functions $\phi \le f \le \psi$ which gave rise to the *upper* and *lower Riemann sums* of f. Continuous functions and monotone functions were seen to be classes of integrable functions. The Riemann integral was shown to be an 'area function', with natural upper and lower bounds and additive over intervals.

Properties of the map $f \mapsto \int_a^b f$ followed from the construction: the map is *linear*, *monotone*, and *continuous*: by the latter we mean that integrable functions which are 'uniformly close' have integrals that remain close to each other. What now remains is to turn this abstract theory into a workable method for *calculation*: and the key to this will be provided by the first result in the next chapter.

FURTHER EXERCISES

1. Use Exercise on 10.2(2) and Proposition 2 to confirm that for integrable functions f the values taken by f at the endpoints of the interval $[a, b]$ do not affect the integral: in particular, show that if $f_{|(a,b)} = c$, where c is a constant, then $\int_a^b f = c(b - a)$.

   (*Hint*: choose $\varepsilon > 0$ and consider $\int_a^b f = \int_a^{a+\varepsilon} f + \int_{a+\varepsilon}^{b-\varepsilon} f + \int_{b-\varepsilon}^b f$. Then let $\varepsilon \to 0$.

2. Suppose that $f : [a, b] \mapsto \mathbb{R}$ is continuous and let $x_k = a + \frac{k(b-a)}{n}$ for $k \le n$. Show that the sequence $(\frac{1}{n} \sum_{k=1}^n f(x_k))_{n \ge 1}$ converges to $f(c)$ for some $c \in [a, b]$.

   (*Hint*: consider the proof of Proposition 2 and use the Mean Value Theorem for Integrals.)

3. (i) Show that if f, g are Riemann-integrable, then so are $\max(f, g)$ and $\min(f, g)$.

   (ii) We saw that if f is integrable then so is $|f|$. Is the converse true? (*Hint*: $f = \mathbf{1}_{\mathbb{Q}} - \mathbf{1}_{\mathbb{R} \backslash \mathbb{Q}}$.)

4. Use Proposition 9.3 to show that, for all x inside its interval of convergence, the power series $\sum_{n \ge 0} \frac{a_n}{n+1} x^{n+1}$ defines a function $F(x) = \sum_{n=0}^{\infty} \frac{a_n}{n+1} x^{n+1}$ which is a primitive of the function $f(x) = \sum_{n=0}^{\infty} a_n x^n$. Apply this to the geometric series $1 - x^2 + x^4 - x^6 + \ldots$ to deduce the formula $1 - \frac{1}{3} + \frac{1}{5} - \frac{1}{7} + \ldots = \frac{\pi}{4}$.

5. The collection $\mathcal{R}(a, b)$ of Riemann-integrable functions on $[a, b]$ and the integral $f \mapsto \int_a^b f$ for $f \in \mathcal{R}(a, b)$ can be defined by a system of axioms. Show that the assumptions stated for $\int_a^b f$ to be an area function (namely that the map is monotone and is additive over intervals) together with the result of Further Exercise 1 above are satisfied for *step functions* if and only if we define $\int_a^b \phi = \sum_{k=1}^n c_k \Delta a_k$ (where $\phi(x) = c_k$ on $(a_{k-1}, a_k]$). Show that this integral is linear.

11 • Integration Techniques

We have developed all the tools we need to show that differentiation and integration really are mutually inverse operations. This fact not only allows us to justify that primitives can provide us with expressions for the ('indefinite') integral, but, using these ideas in combination with the rules for differentiation, we can now justify the various integration techniques one learns in the Calculus. Moreover, the theorems will show where the limits of applicability of these techniques come from. Finally, we can explore integrals of unbounded functions and functions over unbounded intervals by a careful use of limit procedures.

11.1 The Fundamental Theorem of the Calculus

The links between derivatives and integrals form the main theme of this chapter. We prove our results in a form applicable to continuous functions. Recall that for every *integrable* function $f : [a, b] \mapsto \mathbb{R}$, we have already shown in Exercise on 10.3(2) that the map $x \mapsto \int_a^x f$ is a continuous (even Lipschitz-continuous) function from $[a, b]$ to $\mathbb{R}$. We now show that for *continuous f* the map $x \mapsto \int_a^x f$ is actually differentiable, and that its derivative is f.

● Theorem 1 ───────────────────────────

If $f : [a, b] \mapsto \mathbb{R}$ is continuous, then the function $x \mapsto \int_a^x f(t) \, dt$ is a primitive of f. Hence for *any* primitive F of f:

$$\int_a^x f(t) \, dt = F(x) - F(a)$$

PROOF

We apply the Mean Value Theorem for integrals, which we proved at the end of Chapter 10. Write $G(x) = \int_a^x f(t) \, dt$, then

$$G(x + h) - G(x) = \int_x^{x+h} f(t) \, dt = f(x + \theta_h h).h$$

for some $\theta_h \in [0, 1]$. Thus $\frac{G(x+h)-G(x)}{h} = f(x + \theta_h h)$ for each h (recall that we can choose h positive or negative in this version of the Mean Value Theorem for Integrals!). As $h \to 0$, the left-hand side converges to $G'(x)$, while, by continuity of f, the right-hand side converges to $f(x)$. The existence of the limit on the right guarantees the existence of the derivative, and hence $G'(x) = f(x)$. So G is a primitive of f.

Given any primitive F, $x \mapsto F(x) - F(a)$ is also a primitive of f, and we know any two primitives can only differ by a constant. This must therefore apply to the primitives G and $F - F(a)$, which *both* take the value 0 at $x = a$. Hence they are equal everywhere on $[a, b]$. In other words, for all $x \in [a, b]$,

$$F(x) - F(a) = G(x) = \int_a^x f(t)\, dt$$

Example 1

If $f(x) = x^2$ we can now confirm our earlier (laborious!) calculation that $\int_0^1 f = \frac{1}{3}$: since $F(x) = \frac{x^3}{3}$ is a primitive of f, $F(x) - F(0) = \int_0^x f(t)\, dt$ for any x. Hence $\frac{x^3}{3} = \int_0^x f(t)\, dt$ for any x, and in particular: $\int_0^1 f = \frac{1}{3}$.

Example 2

Of course, f need not be continuous in order to be integrable: just as there are many differentiable functions whose derivatives are not continuous! Consider f on $[0, 2]$ given by $f(x) = \begin{cases} 1 & 0 \le x \le 1 \\ 2 & 1 < x \le 2 \end{cases}$ which is discontinuous at 1, but still Riemann-integrable, since it is a step function (recall that the value at endpoints of intervals makes no contribution to the integral). A 'primitive' with domain $[0, 2]$ is given by $F(x) = \begin{cases} x & 0 \le x \le 1 \\ 2x-1 & 1 < x \le 2 \end{cases}$. The choice of F was determined by the need to ensure that it must be continuous, even though it is not really a genuine primitive, since it is not differentiable at 1. However, $F(x) = \int_0^x f$ *does* hold for this continuous function F.

We are now 'justified' in calling any primitive of f an *indefinite integral* $\int f$ of f and we can immediately confirm that power series may be integrated term-by-term within their interval of convergence: we have already seen that $F(x) = \sum_{n=0}^{\infty} \frac{a_n}{n+1} x^{n+1}$ has derivative $f(x) = \sum_{n=0}^{\infty} a_n x^n$ whenever either series converges, so that F is a primitive of f and thus differs from $\int f$ only by a constant. Hence we can finally confirm that $\exp x$ is an indefinite integral of itself, that $\sin$ is an indefinite integral of $\cos$, and many other familiar results of this type. Moreover, the functions defined by power series are infinitely differentiable on their interval of convergence I (that is, they belong to the class $C^\infty(I)$): this follows from the fact that the power series constructed by term-by-term differentiation (and integration) have the same radius of convergence as the original series, so that the arguments can be applied over and over again.

We shall take these immediate consequences of the Fundamental Theorem for granted and concentrate our attention on the various techniques for integration which follow from the rules for differentiation.

Note in passing that integration is a *smoothing* operation: the integral of any integrable function is (Lipschitz-) continuous, even if the original function had many discontinuities. If the original function f is already continuous, then its integral is differentiable, with derivative f. This is in stark contrast to the derivative, since the derivative of a differentiable function need not even be continuous.

EXERCISES ON 11.1

1. Show that if $f \ge g$ on $[a, b]$ then $\int_a^b (f - g)$ represents the area between the graphs of f and g over the interval. Apply this to find the area between $\sin$ and $\cos$ over the interval $[0, 2\pi]$.

2. What is wrong with the following argument? $f(x) = \frac{1}{x^2}$ has $F(x) = -\frac{1}{x}$ as a primitive, so $\int_{-1}^{1} \frac{dx}{x^2} = F(1) - F(-1) = -2$, by the Fundamental Theorem.

11.2 Integration by parts and change of variable

Together with algebraic manipulations such as partial fractions or completing the square the phrases in the title of this section describe the main techniques used in evaluating integrals. They come, respectively, from the product and chain rules for derivatives. The terminology varies between textbooks, but what is often called 'substitution' is really just a version of the change of variable technique. In each of the Propositions below, you should provide a translation to the more 'familiar' Leibniz notation using y for $f(x)$ and $\frac{dy}{dx}$ instead of f'.

● *Proposition 1—Integration by parts*

Suppose that f and g are continuous on $[a, b]$ and that g' exists and is continuous on $[a, b]$. Then:

$$\int_a^b f.g = F(b)g(b) - F(a)g(a) - \int_a^b F.g'$$

for any primitive F of f. In particular, if f' is also continuous on $[a, b]$, then:

$$\int_a^b f'.g = [f.g]_a^b - \int_a^b f.g'$$

where $[f.g]_a^b = f(b)g(b) - f(a)g(a)$.

PROOF
Let H be a primitive of $f.g$, so that for any primitive F of f we have, upon differentiating the function $u = F.g - H$ and using the product rule on $f.g$:

$$u' = (f.g + F.g') - f.g = F.g'$$

In other words, $u = F.g - H$ is a primitive of $F.g'$, so that by the Fundamental Theorem $\int_a^b F.g' = [F.g - H]_a^b$, i.e.

$$\int_a^b F.g' = F(b)g(b) - F(a)g(a) - \{H(b) - H(a)\}$$

which completes the proof, since H is a primitive of $f.g$, so that $H(b) - H(a) = \int_a^b f.g$ as required. The second part follows if we replace f by f' throughout.

● *Example 3*

Term-by-term integration of the power series sin and cos shows that $\int \sin x\, dx = -\cos x$, $\int \cos x\, dx = \sin x$. We can use these to compute $\int_a^b x \cos x\, dx$: set $g(x) = x$, $f(x) = \sin x$, so that $f'(x) = \cos x$. Using the second form of the Proposition, we obtain:

$$\int_a^b x \cos x \, dx = [x \sin x]_a^b - \int_a^b \sin x \, dx$$

In particular, with $[a, b] = [0, \frac{\pi}{2}]$ we have

$$\int_0^{\frac{\pi}{2}} x \cos x \, dx = \frac{\pi}{2} + [\cos x]_0^{\frac{\pi}{2}} = \frac{\pi}{2} - 1$$

Example 4

Using the identity $\cos^2 x + \sin^2 x = 1$ and the Inverse Function Theorem of Chapter 6 it is easy to check that $\sin^{-1}$ has derivative $x \mapsto \frac{1}{\sqrt{1-x^2}}$ on the interval $(-1, 1)$. (But beware! The square root is *undefined* outside $[-1, 1]$ and is 0 at the endpoints. To check that the Inverse Function Theorem applies here you should also verify that $\sin$ is one-one on the interval, so the simple statement 'if $y = \sin^{-1} x$ then $\frac{dy}{dx} = \frac{1}{\sqrt{1-x^2}}$' is not meaningful outside $(-1, 1)$.)

Using integration by parts in the somewhat looser 'indefinite integral' notation, we can now compute $\int \sin^{-1} x \, dx$ using $g = \sin^{-1}$ and $f = 1$. Then

$$\int \sin^{-1} x = x \sin^{-1} x - \int \frac{x}{\sqrt{1 - x^2}} \, dx$$

whenever both sides make sense. To find the last integral we look ahead slightly and 'recognize' $x \mapsto -2x$ as the derivative of $x \mapsto 1 - x^2$, so that with the change of variable $u = 1 - x^2$ we reduce the integral to $-\frac{1}{2} \int u^{-\frac{1}{2}} \, du = u^{\frac{1}{2}}$. Hence $\int \sin^{-1} x \, dx = x \sin^{-1} x + \sqrt{1 - x^2}$ is valid for $x \in (-1, 1)$. However, we still need to justify the effect of the change of variable on the integral.

• Proposition 2—The change of variable formula

Let f be continuous on the closed bounded interval I and suppose that $g : [a, b] \mapsto I$ is continuously differentiable. Then

$$\int_a^b f(g(t)) \cdot g'(t) \, dt = \int_{g(a)}^{g(b)} f(u) \, du$$

PROOF
The function $(f \circ g) \cdot g' : [a, b] \mapsto \mathbb{R}$ is the derivative of $F \circ g$, where F is any primitive of f. To see this, we note simply that by the chain rule, for any $x \in (a, b)$: $(F \circ g)'(x) = F'(y) \cdot g'(x)$ where $y = g(x) \in I$. But then $F'(y) = f(y) = f(g(x))$. So taking $F(x) = \int_a^x f$ as our chosen primitive of f and setting $G(t) = F(g(t))$, we have shown that G is a primitive of $(f \circ g) \cdot g'$ and hence by the Fundamental Theorem:

$$G(b) - G(a) = \int_a^b f(g(t)) \cdot g'(t) \, dt$$

On the other hand, again by the Fundamental Theorem:

$$\int_{g(a)}^{g(b)} f(u) \, du = F(g(b)) - F(g(a)) = G(b) - G(a)$$

which completes the proof. (We should observe, however, that this includes the case where $g(b) < g(a)$, in which our convention $\int_u^v h = -\int_v^u h$ comes into play.)

Example 5

(i) Take $g(t) = t + c$ for a constant c. Then $g' = 1$, so for any continuous $f : [a, b] \mapsto \mathbb{R}$ we obtain:

$$\int_{a+c}^{b+c} f(u) \, du = \int_a^b f(t + c) \, dt$$

(ii) Similarly, when $g(t) = ct$ for a constant c, we obtain $g'(t) = c$ for all t, so that for continuous f on $[a, b]$:

$$\int_{ac}^{bc} f(u) \, du = c \int_a^b f(ct) \, dt$$

(iii) A typical change of variable occurs in:

$$\int \sin^2 t \cos t \, dt = \int u^2 \, du = \frac{u^3}{3} = \frac{1}{3} \sin^3 t$$

where we took $u = \sin t$ and recognized cos as the derivative of sin, so that (formally) $u = g(t) = \sin t$, $g'(t) = \cos t$ and $f(u) = u^2$. (You may fill in appropriate intervals yourself to find values for various definite integrals.)

Example 6

To find $\int_0^1 \frac{1}{1+u^2} \, du$ we take $f(u) = \frac{1}{1+u^2}$ and $g(t) = \tan t$, so that $g'(t) = \sec^2 t$. Here $g([0, \frac{\pi}{4}]) = [0, 1]$, and g is increasing, so that

$$\int_0^1 \frac{du}{1 + u^2} = \int_0^{\frac{\pi}{4}} \frac{1}{1 + \tan^2 t} \cdot \sec^2 t \, dt = \int_0^{\frac{\pi}{4}} dt = \frac{\pi}{4}$$

The change of variable formula is frequently called 'integration by substitution': this merely involves a shift of emphasis, such as we have used in Example 6: whereas in Example 5 our concern was to simplify the integrand by 'spotting' derivatives, we introduced a substitution of variables in Example 6 which had the effect of simplifying the integral after accounting for the derivative (i.e. g'). Thus the formula remains the same, but attention shifts from integrating f over the interval $[\alpha, \beta] = [g(a), g(b)]$ to integrating f over $[a, b]$. In Example 6, $g = \tan$ and $[a, b] = [0, \frac{\pi}{4}]$, so that $[\alpha, \beta] = [0, 1]$.

If f is continuous and strictly increasing on $[a, b]$, so that $f([a, b]) = [f(a), f(b)]$ is the image of $[a, b]$, then f^{-1} is continuous, and we can use integration by parts on:

$$\int_a^b f(x) \, dx = \int_a^b 1.f(x) \, dx = [xf(x)]_a^b - \int_a^b xf'(x) \, dx$$

But $\int_a^b xf'(x)\,dx = \int_{f(a)}^{f(b)} f^{-1}(y)\,dy$ if we make the substitution $y = f(x)$ and apply the change of variable formula with f in place of g and f^{-1} in place of f. Hence:

$$\int_a^b f + \int_{f(a)}^{f(b)} f^{-1} = bf(b) - af(a)$$

TUTORIAL PROBLEM 11.1

Give a simple graphical representation of this identity. By differentiating the primitives $\int_a^x f + \int_{f(a)}^{f(x)} f^{-1}$ and the function $x \mapsto xf(x)$ directly, find an alternative proof of the above result.

Example 7

The integral $\int_{-1}^1 \sqrt{1 - x^2}\,dx$ can be evaluated using the substitution $x = g(t) = \sin t$ on the interval $[-\frac{\pi}{2}, \frac{\pi}{2}]$ (which ensures also that g is one-one), so that $g'(t) = \cos t$ is continuous on $[-1, 1] = g([-\frac{\pi}{2}, \frac{\pi}{2}])$. Hence we obtain

$$\int_{-1}^1 \sqrt{1 - x^2}\,dx = \int_{-\frac{\pi}{2}}^{\frac{\pi}{2}} \sqrt{1 - \sin^2 t}\,.\cos t\,dt = \int_{-\frac{\pi}{2}}^{\frac{\pi}{2}} \cos^2 t\,dt = \int_{-\frac{\pi}{2}}^{\frac{\pi}{2}} \frac{1}{2}(1 + \cos 2t)\,dt$$

which has $F : t \mapsto \frac{t}{2} + \frac{1}{4}\sin 2t$ as a primitive. Hence the integral reduces to $F(\frac{\pi}{2}) - F(-\frac{\pi}{2}) = \frac{\pi}{2}$.

EXERCISES ON 11.2

1. Use integration by parts to find $\int_{e-1}^e \log x\,dx$.
2. Show that if $f : I \mapsto \mathbb{R}$ is continuous then

$$\int_a^b t^{n-1} f(t^n)\,dt = \frac{1}{n}\int_{a^n}^{b^n} f(x)\,dx$$

3. Recall from Tutorial Problem 7.2 in Chapter 7 that $\frac{\pi}{2}$ can be defined (without recourse to geometry!) as the unique point ζ in $(0, 2)$ at which cos takes the value 0. Geometrically, of course, π is defined as the ratio of the circumference of a circle to its diameter, and this leads to the formula $A = \pi r^2$ which links the area and radius. Calculate $\int_0^1 \sqrt{1 - x^2}$, using the analytic definition $\pi = 2\zeta$, and show that this coincides with the geometric definition of π.

11.3 Improper integrals

So far we have taken care to integrate only bounded functions over bounded intervals. We now use limit operations to extend this in both directions – the results will still be denoted by integrals, but for our definitions they can no longer be described as 'proper' Riemann integrals, since they are not defined in terms of upper and lower Riemann sums. They are thus traditionally known as *improper Riemann integrals*, though the term 'singular integral' might be more appropriate, as they deal with situations where a function has a singularity, i.e. the area under the graph is not described by any 'closed curve', even though it will (in many situations) be approximated by finite areas of that type.

Thus there are three types of improper integrals:

(i) where a bounded function is integrated over an unbounded interval; e.g. $\int_{-\infty}^{\infty} \frac{1}{1+x^2} \, dx$, or $\int_0^{\infty} e^{-x} \, dx$.

(ii) where an unbounded function is integrated over a bounded interval; e.g. $\int_0^1 \frac{dx}{\sqrt{x}}$.

(iii) where an unbounded function is integrated over an unbounded interval; e.g. $\int_1^{\infty} \frac{dx}{x^2}$.

We define *convergent* integrals of each of the above types. If an integral is not convergent, we call it *divergent*.

Suppose f is bounded on $[a, \infty)$ and integrable on $[a, b]$ for each $b > a$. If $\lim_{b \to \infty} \int_a^b f(x) \, dx$ exists we write $\int_a^{\infty} f(x) \, dx$ for this limit and say that the integral *converges*. The improper integral $\int_{-\infty}^b f(x) \, dx = \lim_{a \to -\infty} \int_a^b f(x) \, dx$ is defined similarly, and we set

$$\int_{\mathbb{R}} f(x) \, dx = \int_{-\infty}^{\infty} f(x) \, dx = \int_{-\infty}^0 f(x) \, dx + \int_0^{\infty} f(x) \, dx$$

If the real function f is defined on $(a, b]$ and integrable on $[a + \varepsilon, b]$ whenever $0 < \varepsilon < b - a$, then $\int_a^b f(x) \, dx$ is defined as $\lim_{\varepsilon \downarrow 0} \int_{a+\varepsilon}^b f(x) \, dx$ whenever this limit exists. Similarly, $\int_a^b f(x) \, dx = \lim_{\varepsilon \downarrow 0} \int_a^{b-\varepsilon} f(x) \, dx$ when this limit exists.

Improper integrals of types (i), (ii) and (iii) are sometimes known as integrals of *first, second* or *third kind* respectively (though we should not take the analogy with extraterrestrials too far!). All three cases have some striking features in common with series.

The sequence of *partial integrals* $\int_a^b f(t) \, dt$ ($b \le \infty$) plays the same role as do the partial sums $F_n = \sum_{k=0}^n f_k$ of a series $\sum_{k \ge 0} f_k$: setting $F(x) = \int_a^x f(t) \, dt$ we have a convergent integral $\int_a^b f(t) \, dt$ if and only if the function F converges as $x \to b$. Conversely, suppose that F is a given, continuously differentiable function. Then $F(x) = F(a) + \int_a^x F'(t) \, dt$, and the improper integral of F' exists if and only if F converges as $x \to b$.

⊛ *Example 8*

For fixed $\alpha \in \mathbb{R}$, $\int_1^{\infty} \frac{dt}{t^{\alpha}}$ has $f(t) = t^{-\alpha}$ continuous on each interval $[1, x]$ ($x \ge 1$) and for $\alpha \ne 1$ we have $\int_1^x \frac{dt}{t^{\alpha}} = \frac{1}{-\alpha+1}(x^{-\alpha+1} - 1)$, which converges to $\frac{1}{\alpha-1}$ as $x \to \infty$, whenever $\alpha > 1$. When $\alpha < 1$ the integral diverges, and when $\alpha = 1$, the function $f(t) = \frac{1}{t}$ has primitive log, as we have seen previously. But $\log x \to \infty$ as $x \to \infty$, so that the improper integral $\int_1^{\infty} \frac{dt}{t^{\alpha}} = \frac{1}{\alpha-1}$ exists *if and only if* $\alpha > 1$.

Next, consider $\int_0^1 \frac{dt}{t^{\alpha}}$. This is an improper integral of the second kind, since $t^{-\alpha}$ is unbounded near 0. Now we obtain: $\int_{\varepsilon}^1 \frac{dt}{t^{\alpha}} = \frac{1}{-\alpha+1}(1 - \varepsilon^{-\alpha+1})$, which converges to $\frac{1}{1-\alpha}$ as $\varepsilon \downarrow 0$ if and only if $\alpha < 1$. Hence $\int_0^1 \frac{dt}{t^{\alpha}} = \frac{1}{1-\alpha}$ when $\alpha < 1$, and diverges otherwise.

Putting these pieces together, we have shown that the improper integral $\int_0^{\infty} \frac{dt}{t^{\alpha}}$ *diverges for all real* α.

1. Evaluate the following improper integrals if they converge:

   (i) $\int_0^1 \log x \, dx$ (ii) $\int_{-1}^1 \frac{dx}{\sqrt{1-x^2}}$ (iii) $\int_{-\infty}^\infty \frac{dx}{e^x + e^{-x}}$

2. For which a, b does the integral $\int_a^b \frac{dx}{x(x-1)}$ exist? Evaluate it in terms of a and b where possible.

3. Show that for any $a \in \mathbb{R}$ the map $f \mapsto \int_a^\infty f$ is linear, i.e. that if $\int_a^\infty f$ and $\int_a^\infty g$ converge, then for any $\alpha, \beta \in \mathbb{R}$ we have:

$$\int_a^\infty (\alpha f + \beta g) = \alpha \int_a^\infty f + \beta \int_a^\infty g$$

11.4 Convergent integrals and convergent series

We can regard real series as a special case of improper integrals: given a series $\sum_{k \geq 0} f_k$, with $f_k \in \mathbb{R}$ for all $k \geq 0$, define the function $f : [0, \infty) \mapsto \mathbb{R}$ by setting $f(t) = f_k$ on the interval $(k-1, k)$ for each $k \geq 1$ (and arbitrarily at integer points, since the endpoints of intervals do not affect the integral).

TUTORIAL PROBLEM 11.2

> Prove carefully that $\int_0^\infty f(t) \, dt = \sum_{k=0}^\infty f_k$ whenever either side converges, and that if $f(x) \geq 0$ for all x, then $\int_0^\infty f(t) \, dt$ converges if and only if it is bounded.

This analogy between series and integrals also extends to a

• Comparison Test

If $0 \leq f(x) \leq g(x)$ for all $x \geq a$, and if $\int_a^b g(t) \, dt$ is convergent, then so is $\int_a^b f(t) \, dt$. If $\int_a^b f(t) \, dt$ diverges, so does $\int_a^b g(t) \, dt$.

PROOF
This follows at once from the inequalities

$$\int_a^x f \leq \int_a^x g \leq \int_a^b g$$

which hold for all $x \in [a, b]$; the first since the Riemann integral is monotone, and the second since the integral is additive and since $\int_x^b g \geq 0$ when $g \geq 0$. Thus $\lim_{x \to b} F(x) \leq \int_a^b g$, where $F(x) = \int_a^x f$.

• Example 9

The integral $\int_0^\infty \frac{dt}{1+t^2}$ is easily seen to converge, since $g(t) = \frac{1}{1+t^2}$ has primitive $\tan^{-1}$, so that $\int_0^b \frac{dt}{1+t^2} = \tan^{-1} b$ for all $b > 0$. But $\tan^{-1} b \to \frac{\pi}{2}$ as $b \to \infty$. On the other hand, since g is an even function of t, $\int_\mathbb{R} \frac{dt}{1+t^2} = 2 \int_0^\infty \frac{dt}{1+t^2} = \pi$. Let

$f(t) = \frac{e^{-t}}{1+t^2} = e^{-t}g(t)$. Then $0 \leq f \leq g$ on $(0, \infty)$, so that by the Comparison Test we see that $\int_0^\infty \frac{e^{-t}}{1+t^2}\, dt$ converges.

⊕ *Example 10*

The evaluation of $\int_0^4 \frac{dt}{(t-1)^{\frac{1}{3}}}$ causes a new problem: the integrand $f(t) = (t-1)^{-\frac{1}{3}}$ is *undefined* at $t = 1$. We can hope to overcome this by first finding the sum of the integrals $\int_0^{1-\varepsilon} f$ and $\int_{1+\varepsilon}^4 f$ and then hoping for a limit as $\varepsilon \downarrow 0$. In this case the ploy works, and the limit turns out to be $\int_0^4 f(t)\, dt = \frac{3}{2}(9\sqrt{3} - 1)$. However, this technique also fails frequently, since we have no guarantee that the limit will exist: for example, $\int_0^4 (1 - t^2)^{-2} dt$ diverges. Note that a careless 'application' of the Fundamental Theorem would give the value $\frac{4}{3}(!)$.

Our next Proposition exploits the analogy with series further, providing a proof of the Integral Test for series, which was already stated in the long 'digression' in Section 3.3. We can state it slightly more precisely.

● *Proposition 3*

If $f \geq 0$ is monotone decreasing, then the sequence (a_n) defined by $a_n = \sum_{k=1}^n f(k) - \int_1^{n+1} f(t)\, dt$ is non-negative, monotone increasing and convergent, and $a = \lim_{n\to\infty} a_n$ satisfies $0 \leq a \leq f(1)$. Therefore the improper integral $\int_1^\infty f(t)\, dt$ converges if and only if the series $\sum_{k\geq 1} f(k)$ converges.

PROOF
As the integral is monotone, $f(k+1) \leq \int_k^{k+1} f(t)\, dt \leq f(k)$, and summing this for $k = 1, \ldots, n$ we obtain $\sum_{k=2}^{n+1} f(k) \leq \int_1^{n+1} f(t)\, dt \leq \sum_{k=1}^n f(k)$. This already shows that the partial sums of the series $\sum_k f(k)$ will converge if and only if the integrals do. Moreover, the sequence (a_n), as defined above, is non-negative and $a_n < f(1) - f(n+1)$, by the second string of inequalities. As $a_n - a_{n-1} = f(n) - \int_n^{n+1} f(t)dt \geq 0$, the sequence (a_n) is non-decreasing and bounded above, and so converges. Clearly $a \leq f(1) - \lim_{n\to\infty} f(n+1)$.

Note that if, moreover, $\lim_{t\to\infty} f(t) = 0$ then we also have:

$$a = \lim_{n\to\infty} a_n = \lim_{n\to\infty} \left\{ \sum_{k=1}^n f(k) - \int_1^n f(t)\, dt \right\} = \sum_{k=1}^\infty f(k) - \int_1^\infty f(t)\, dt$$

so that a gives a precise value of the difference between the 'area under the graph' of f and its approximation by the 'partition' $\{1, 2, 3, \ldots, n, \ldots\}$.

TUTORIAL PROBLEM 11.3

These ideas were exploited in Chapter 3 to define *Euler's constant* γ. We can also use them for the *Riemann zeta function* $\zeta(\alpha) = \sum_{k=1}^\infty k^{-\alpha}$, which plays a major role in advanced number theory. Use the above inequalities with $f(t) = t^{-\alpha}$ to derive the (strict!) inequalities for $\alpha > 1$:

$$\frac{1}{\alpha - 1} < \zeta(\alpha) < \frac{\alpha}{\alpha - 1}$$

and conclude that $\lim_{\alpha \to \infty} \zeta(\alpha) = 1$.

● *Example 11*

We can also use the Integral Test to distinguish between the series $\sum_{n \geq 2} \frac{1}{n \log n}$ and $\sum_{n \geq 2} \frac{1}{n(\log n)^\alpha}$ for $\alpha > 1$: we compare the behaviour of the first with that of the integral $\int_2^\infty \frac{dx}{x \log x}$, which diverges, since

$$\int_2^b \frac{dx}{x \log x} = \int_{\log 2}^{\log b} \frac{du}{u} = \log(\log b) - \log(\log 2)$$

which is unbounded above. On the other hand, the integral $\int_2^\infty \frac{dx}{x(\log x)^\alpha}$ converges, since

$$\int_2^b \frac{dx}{x(\log x)^\alpha} = \int_{\log 2}^{\log b} \frac{du}{u^\alpha}$$

$$= \frac{1}{-\alpha + 1}\{(\log b)^{-\alpha+1} - (\log 2)^{-\alpha+1}\} \to \frac{1}{\alpha - 1}(\log 2)^{1-\alpha}$$

as $b \to \infty$.

Hence $\sum_{n \geq 2} \frac{1}{n \log n}$ diverges, while $\sum_{n \geq 2} \frac{1}{n(\log n)^\alpha}$ converges for every $\alpha > 1$.

Just as for series, we define the improper integral $\int_a^\infty f$ to be *absolutely convergent* if $\int_a^\infty |f|$ converges. The integral $\int_a^\infty f$ is *conditionally convergent* if it converges, but does not converge absolutely.

We leave it for the Exercises below to show that absolutely convergent integrals converge. Examples of such integrals include $I_\alpha = \int_\pi^\infty \frac{\sin x}{x^\alpha} dx$ for $\alpha > 1$. This is most easily seen by comparing $J_\alpha = \int_\pi^\infty \frac{|\sin x|}{x^\alpha} dx$ with the convergent integral $\int_\pi^\infty \frac{dx}{x^\alpha}$, since $|\sin x| \leq 1$.

However, this does not extend to the case $0 < \alpha \leq 1$, where the same integrand yields a simple example of a conditionally convergent integral: now we have, for $K > 1$,

$$\int_\pi^{K\pi} \frac{|\sin x|}{x^\alpha} dx = \sum_{k=1}^{K-1} \int_{k\pi}^{(k+1)\pi} \frac{|\sin x|}{x^\alpha} dx \geq \sum_{k=1}^{K-1} \int_{(k+\frac{1}{4})\pi}^{(k+\frac{3}{4})\pi} \frac{|\sin x|}{x^\alpha} dx$$

$$\geq \sum_{k=1}^{K} \frac{\pi}{2}\left\{ \frac{1}{\sqrt{2}(k + \frac{3}{4})^\alpha} \right\}$$

since $|\sin x| \geq \frac{1}{\sqrt{2}}$ and $x^\alpha \leq (k + \frac{3}{4})^\alpha$ on each subinterval. But since the series $\sum_{n \geq 1} n^{-\alpha}$ diverges when $0 < \alpha \leq 1$, so does the integral, so J_α is divergent.

To see that I_α converges for $0 < \alpha \leq 1$, we need only integrate by parts to obtain:

$$\int_\pi^b \frac{\sin x}{x^\alpha} dx = \left[\frac{-\cos x}{x^\alpha} \right]_\pi^b - \alpha \int_\pi^b \frac{\cos x}{x^{\alpha+1}} dx$$

and the last integral converges absolutely, by comparison with $\int_\pi^b \frac{dx}{x^\beta}$ for $\beta = \alpha + 1 > 1$. Hence I_α converges conditionally.

EXERCISES ON 11.4

1. State and prove a comparison test for improper integrals of the second kind.
2. Use the Integral Test to check that $\sum_{n \geq 1} n^{-\alpha}$ converges if and only if $\alpha > 1$.
3. Show that absolute convergence implies convergence for improper integrals, and that $|\int_a^\infty f| \leq \int_a^\infty |f|$ whenever the latter is finite. (*Hint*: for the first part look up the proof for series in Chapter 3, and use the Comparison Test for integrals.)

Summary

This chapter rounded off our introduction to Analysis with the main result (the Fundamental Theorem) linking the two parts of the Calculus. We then made use of primitives and the rules for differentiation to derive some simple techniques for transforming integrals into each other, always in the hope that the transformed version will have a simple solution. We also explored how one can combine integrals with limit operations, in order to handle unbounded functions and integrate over unbounded intervals. Here there is a simple analogy with series, which we used to derive theorems on the comparison of improper integrals, and to define the notion of absolute convergence for integrals. Conversely, the properties of integrals allowed us to prove the Integral Test for series, which has been used already in Chapter 3 to discuss Euler's constant and the behaviour of the harmonic series.

FURTHER EXERCISES

1. Criticize the following calculation: Let $I = \int_{-1}^1 \frac{dx}{1+x^2}$. Substitute $x = \frac{1}{u}$, so that $dx = -\frac{du}{u^2}$ and $I = \int_{-1}^1 \frac{-du}{u^2+(1+1/u^2)} = -I$, which means that $I = 0$.

   First give an obvious reason why this must be wrong, then explain the mistake and calculate the correct value of I.
2. For which a, b do the following integrals exist? When they do, evaluate the integrals in terms of a and b.
   (i) $\int_a^b \frac{dx}{x^2-1}$ (ii) $\int_a^b \tan x \, dx$ (iii) $\int_a^b \sin^{-1} x \, dx$.
3. Find a primitive for $f(x) = \exp(|x|)$ over the whole of $\mathbb{R}$. (*Hint*: you can discuss the intervals $[0, \infty)$ and $(-\infty, 0]$ separately, but remember that the two 'pieces' must fit together in a differentiable way at 0.)
4. Investigate the convergence of the series $\sum_{n \geq 3} \frac{1}{n \log n \log(\log n)^\alpha}$ for $\alpha \geq 1$.
5. Prove the *second mean value theorem* for integrals: If $f > 0$ and $f' < 0$ are both continuous on $[a, b]$ and if $\int_a^c g \in [m, M]$ for all $c \in [a, b]$, then $mf(a) \leq \int_a^b fg \leq Mf(a)$. Hence show that $\int_a^b fg = f(a) \int_a^d g$ for some $d \in (a, b)$.
6. Use the change of variable $x = t^2$ to analyse the *Fresnel integral*, by showing that $\int_0^\infty \sin t^2 \, dt = \frac{1}{2} \lim_{u \to \infty} \int_0^{u^2} \frac{\sin u}{\sqrt{u}} \, du$. Now construct an alternating series by

considering the last integrand at points $k\pi$ for $k \geq 0$, and use this to explain why the first integral converges, even though the integrand does not tend to 0.

7. Explain why $\Gamma(x) = \int_0^\infty t^{x-1}e^{-1}dt$ defines a convergent improper integral of the third kind when $x > 0$. Calculate $\Gamma(n)$ for $n \in \mathbb{N}$. Show that $\Gamma(x+1) = x\Gamma(x)$ for all $x > 0$, so that the *gamma function* Γ provides a way of extending the *factorial n!* from $\mathbb{N}$ to $(0, \infty)$. For a brief discussion of some of its main properties, and the sense in which this is the *only* such extension of the factorial, you may consult K.G. Binmore: *Mathematical Analysis*, Chapter 17.

12 • What Next? Extensions and Developments

We have only scraped the surface of many topics in Real Analysis – and said nothing at all about the beautiful and very important subject of *Complex Analysis*, that is, the analysis of functions defined on the complex plane $\mathbb{C}$. A fairly recent text which provides a detailed introduction to the latter is *Basic Complex Analysis* by J.E. Marsden and M.J. Hoffman (2nd edition), WH Freeman, 1987.

In this final chapter we provide a glimpse of some of the directions into which the ideas of Analysis have been extended in the past 150 years. As the mathematical complexity increases, the record of these more recent developments is rather less digestible than that of earlier centuries. However, the overview text by Dirk Struik, *A Concise History of Mathematics* (Dover) provides a useful perspective, while the rather more detailed text by C.B. Boyer and U. Merzbach, *A History of Mathematics* (2nd edition), John Wiley and Sons, 1989, describes the mathematics in somewhat greater depth and provides many guides to further reading.

12.1 Generalizations of completeness

Analysis in *n* dimensions and beyond

Much of Analysis is concerned with conditions under which our intuitive 'feel' for the behaviour of functions can be justified, and with exploring under what conditions results can be extended to a more general framework. We saw, for example, that the *completeness* of the real line plays a crucial role in establishing most of the important results we have proved. To what extent can this idea be captured in higher dimensions, that is, for functions defined in the plane $\mathbb{R}^2$ or more generally in the *n*-dimensional space $\mathbb{R}^n$? Here we need to be clear first that $\mathbb{R}^n$ consists of *n*-dimensional *vectors* $\mathbf{x} = (x_1, x_2, ..., x_n)$ whose coordinates (x_i) are given relative to some basis – we assume this to be the standard basis described, for example, in R.B.J.T. Allenby's *Linear Algebra* in this series (Modular Mathematics). However, there are two immediate problems:

(i) what should play the (crucial) role of the *modulus* $|.|$ as 'distance function' between vectors?

(ii) how do we handle the problem that $\mathbb{R}^n$ is not linearly ordered; for example, how would we tell which of the vectors $\mathbf{x} = (1, 2)$ and $\mathbf{y} = (2, 1)$ is the 'bigger'? Therefore we can no longer use the supremum of sets that are bounded above in the same way as in $\mathbb{R}$. What takes its place?

The first question has a natural answer when we consider straight lines as the 'shortest distance between two points' (which, of course, assumes that our space is 'flat', as did Euclid for his two- and three-dimensional geometry!): $|x| = \sqrt{x^2}$ can be replaced by $\|\mathbf{x}\|_2 = \sqrt{x_1^2 + x_2^2}$ when $\mathbf{x} = (x_1, x_2) \in \mathbb{R}^2$, and by $\|\mathbf{x}\|_2 = \sqrt{x_1^2 + x_2^2 + x_3^2 + \ldots + x_n^2}$ when $\mathbf{x} = (x_1, x_2, \ldots, x_n) \in \mathbb{R}^n$. This *Euclidean norm* of the vector $\mathbf{x}$ has all the properties of 'length' that we require, and $\|\mathbf{x} - \mathbf{y}\|_2$ then represents the distance between the vectors $\mathbf{x}$ and $\mathbf{y}$.

For the second question we have to change our perspective a little: the property of 'having no gaps' can also be stated in terms of sequences of vectors as follows: if the distances $\|\mathbf{x}_k - \mathbf{x}_m\|_2$ between the elements of a sequence in $\mathbb{R}^n$ become arbitrarily small, then we would expect these elements 'eventually' to get closer and closer to some fixed element $\mathbf{x} \in \mathbb{R}^n$. In other words, if $(\mathbf{x}_k)$ has the property that $\|\mathbf{x}_k - \mathbf{x}_m\|_2 \to 0$ as $k, m \to \infty$ (we say that $(\mathbf{x}_k)$ is a *Cauchy sequence*) then it should converge to some $\mathbf{x} \in \mathbb{R}^n$, i.e. $\|\mathbf{x}_k - \mathbf{x}\|_2 \to 0$ as $k \to \infty$.

In short: completeness means that all Cauchy sequences converge. (You should now prove that in $\mathbb{R}$ this is equivalent to our definition, using the Bolzano–Weierstrass Theorem!) This theorem and its consequences (e.g. the concept of accumulation points) holds the key to the validity of many of our theorems in higher dimensions.

The Euclidean norm is not the only idea on 'length' we could impose on $\mathbb{R}^n$. Remarkably, however, all reasonable *norms* on $\mathbb{R}^n$ turn out to be essentially the same! To understand this, we need to explore the roles of open intervals on $\mathbb{R}$ further, since these were precisely what we meant by the *neighbourhood* of a given point: in $\mathbb{R}^n$ these are replaced by 'open balls' of the type $B(\mathbf{a}, \delta) = \{\mathbf{y} \in \mathbb{R}^n : \|\mathbf{x} - \mathbf{y}\|_2 < \delta\}$ and a set A in $\mathbb{R}^n$ is *open* if for every point $\mathbf{a} \in A$ we can find $\delta > 0$ such that the open ball $B(\mathbf{a}, \delta) \subset A$. A set is *closed* if its complement is open: it turns out that these are precisely the sets which contain all their accumulation points. This is the starting point for *topological* considerations in Analysis.

The *derivative* also takes on a new guise in $\mathbb{R}^n$: we cannot define f' directly as the lim of a ratio, since ratios of vectors are not meaningful. There are a number of ways round this dilemma: the most effective is to recognize that the derivative $f'(a)$ provides an *affine approximation* to the function f at the given point $\mathbf{a}$. We saw this already in Chapter 8 in our discussion of the 'chord-slope function'. In $\mathbb{R}^n$ we can use this idea to *define* the value of the derivative not as a number, but as an affine mapping. This, again, provides close links between analysis and linear algebra.

Much of the above, and its generalizations to more abstract *metric spaces*, can be found, for example, in Marsden and Hoffman's *Elementary Classical Analysis* (2nd edition, also published by WH Freeman). For a more general approach and introduction to modern 'functional analysis', which combines abstract algebraic ideas with topology and analysis, see the classics by W.H. Rudin: *Principles of Mathematical Analysis, Real and Complex Analysis* and *Functional Analysis*. These three texts provide a wealth of information on all the questions discussed in this chapter.

12.2 Approximation of functions

Much of our discussion of power series (and especially Taylor expansions) centred on the idea that fairly complicated functions could be approximated arbitrarily closely by simpler ones, such as polynomials. In fact, we have all the information needed to prove a famous theorem due to Weierstrass: Every continuous function $f : [0, 1] \mapsto \mathbb{R}$ can be uniformly approximated by polynomials. This simply means that for given $\varepsilon > 0$ there exists a polynomial p such that $|f(x) - p(x)| < \varepsilon$ for all $x \in [0, 1]$. In fact, a sequence of such polynomials was explicitly constructed by Sergei Bernstein (1880–1968):

$$p_n(x) = \sum_{k=0}^{n} \binom{n}{k} f\left(\frac{k}{n}\right) x^k (1 - x)^{n-k}$$

Approximation of general functions by functions taken from a 'well-known' class, such as polynomials or trigonometric functions, forms a further topic in which much success was achieved in the nineteenth and early twentieth century. *Fourier series* are an outstanding example of the success of analytic ideas in applications to the solution of partial differential equations, the modelling of waves and many other physical phenomena, as well as in providing the trigger for many counter-examples in the development of Analysis itself – the example produced by Abel, which we discussed at the end of Chapter 7, was only the first of many!

12.3 Integrals of real functions: yet more completeness

Our discussion of the Riemann integral alluded only briefly to *Lebesgue's criterion* for integrability. In fact, this is at the heart of another major extension of Analysis which was carried out at the turn of the century, and led to the identification of various *norms* on infinite-dimensional vector spaces, enabling us, in particular, to discuss the concept of the 'length' of a function in several essentially different ways – this is in complete contrast to what happens in $\mathbb{R}^n$.

Recall from Chapter 10 that if a sequence (f_n) of real functions converges *pointwise* to the function f, then the integrals $\int_0^1 f_n$ need not converge to $\int_0^1 f$. To avoid problems of this kind, we introduced the idea of *uniform* convergence, by giving the vector space $C[0, 1]$ of all continuous functions on $[0, 1]$ the *uniform norm*, defined by $\|f\|_\infty = \sup\{|f(x)| : 0 \leq x \leq 1\}$. This enabled us to show that if a sequence $\{f_n\}$ in $C[0, 1]$ converges uniformly to f (which just means that $\|f_n - f\|_\infty \to 0$ as $n \to \infty$), then also

$$\int_0^1 f_n(x) \, \mathrm{d}x \to \int_0^1 f(x) \, \mathrm{d}x$$

But the 'distance' $\|f - g\|_\infty$ has nothing to do with integration as such, and no similarly nice results can be obtained for $C[0, 1]$ when it is given the more 'natural' norm $\|f\|_1 = \int_0^1 |f(x)|\,dx$ instead: in fact, defining

$$f_n(x) = \begin{cases} 0 & \text{if } 0 \le x \le \frac{1}{2} \\ n(x - \frac{1}{2}) & \text{if } \frac{1}{2} < x < \frac{1}{2} + \frac{1}{n} \\ 1 & \text{otherwise} \end{cases}$$

we can show that $\|f_n - f_m\|_1 \to 0$ as $m, n \to \infty$ (so that $\{f_n\}$ is a *Cauchy sequence* in this norm), and yet there is no continuous function f to which this sequence converges, i.e. there is no f in $C[0, 1]$ such that $\|f_n - f\|_1 \to 0$ as $n \to \infty$. (Sketching the area between f_n and f_m for $m < n$ should convince you that the sequence is Cauchy. Next, suppose $\|f_n - f\|_1 \to 0$, and show that we must have $\int_0^{\frac{1}{2}} |f(x)|\,dx = 0$, while on the other hand, $f(x) = 1$ if $x > \frac{1}{2}$. Explain why this is impossible when f is continuous.)

Thus, as a normed space, $(C[0, 1], \|.\|_1)$ is *not* 'complete': we have found a Cauchy sequence which fails to have a limit in this space. This is rather similar to the situation which led us to work with the set $\mathbb{R}$ rather than with $\mathbb{Q}$. Recalling the crucial importance of completeness in the case of $\mathbb{R}$, we naturally look for a theory of integration which does not have this shortcoming. This is precisely what was achieved by Lebesgue's theory, which includes the Riemann integral as a special case, and also solves the other problems listed.

Central to this theory is the concept of the *measure* of a set, which generalizes the idea of 'length' from intervals to more general sets. This, in turn, necessitates a much closer scrutiny of the structure of the continuum, $\mathbb{R}$, which is exactly where we started from! Therefore this seems like a good place to stop.

Appendix A: Program Listings

The following contains a number of listings of code fragments from Pascal programs as mentioned in various Tutorial Problems. These need to be completed to make them into full Pascal programs, e.g. add **begin end** statments around blocks, add variable initialization, and initialize graphics — you will need to find out how to use graphics on your particular system. The program fragments can be easily altered to work in other programming languages.

A.1 Sequences program

The following program illustrates the behaviour of $x^2 - c$, for a given constant c (*See* Chapter 3 and Section 2 in Chapter 7.)

```
function phi(x:real):real;
 phi:=x*x-c;

procedure list;{Numerical listing: lists the terms in the sequence}
writeln('x[ 0 ] = ',guess);
for n:=1 to maxn do
  guess:=phi(guess);
  writeln('x[',N:3,'] = ',guess:10:7);

procedure plot;{draws the graph of the saw-tooth sequence}
 moveto((0.0),(guess));
 for n:=1 to maxn do
   guess:=phi(guess);
   IF (Ymin<guess) AND (ymax>guess) THEN lineto((n),(guess));

procedure spider;{draws the graphs of y=phi(x) and y=x and the SPIDER}
 MOVETO((guess),(0.0));
 for n:=1 to maxn do
  y:=phi(guess);
  lineto((guess),(y));
  lineto((y),(y));
  guess:=y;

begin{main program block }
readln('Input value of constant c in  x^2-c');readln(c);
write('initial guess');readln(guess);
write('number of iterations'); readln(maxn);
list;plot;spider;
end.
```

A.2 Another sequence program

This program is very similar to the previous one, but looks at a variety of sequences; see Tutorial Problem 3.3.

```
FUNCTION POWER(a,b :real) :REAL;
 power:=exp(b*ln(a));

FUNCTION FAC(M:INTEGER) : real;
 facJ:=1;
 FOR I:=1 TO N DO facJ:=facJ*I;
 FAC:=facJ;

PROCEDURE GENERALTERM;
 IF Tipe=1 THEN
  WRITELN('Input value of k');  readln(k);
  a[0]:=0;
  FOR N:=1 to MAXN DO a[n]:=POWER(n,k)/POWER(1+1/k,n);

 IF Tipe=2 THEN
  WRITELN('Input value of x in (0,1]');  readln(x);
  a[0]:=1;
  FOR N:=1 to MAXN DO a[n]:=power(x,1/n);

 IF Tipe=3 THEN
  a[0]:=1;
  FOR N:=1 to MAXN DO a[n]:=power(n,1/n);

 IF Tipe=4 THEN
  WRITELN('Input value of x,|x|<1,and p,p>=0'); readln(x,p);
  a[0]:=0;
  FOR N:=1 to MAXN DO a[n]:=power(n,p)*power(x,n);

 IF Tipe=5 THEN
  WRITELN('Input value of p');  readln(p);
  a[0]:=0;
  FOR N:=1 to MAXN DO a[n]:=power(n,p)/fac(n);

 IF Tipe=6 THEN
  a[0]:=1;
  FOR N:=1 to MAXN DO a[n]:=COS(N)/N;

procedure list;{lists the terms in the sequence}
 xmin:=0;xmax:=maxn;ymin:=0;ymax:=0;
 writeln('a[ 0 ] = ',a[0]:15:7);
 for n:=1 to maxn do
  if a[n]>ymax then ymax:=a[n];
  if a[n]<ymin then ymin:=a[n];
  writeln('a[',N:3,'] = ',a[n]:15:7);
 WRITELN('The minimum and maximum values are ',ymin:12:5,'  and
 ',ymax:12:5);

procedure plot;{draws the graph of the saw-tooth sequence}
 moveto(scalex(1.0),scaley(a[1]));
 for n:=2 to maxn do  lineto(scalex(n),scaley(a[n]));

begin{main program block }
 WRITELN('Which sequence do you want:');READLN(Tipe);
 WRITE('Input number of terms');
 IF (Tipe=5) THEN
  begin WRITELN('   : N.B.  N < 30 to avoid overflow with n!');
```

$2 \leq x_k \leq 3$ implies $2 \leq x_{k+1} \leq 3$, so by induction all x_n lie between 2 and 3. Next, we show that (x_n) is decreasing:

$$x_{n+1} - x_n = \frac{1}{5}(x_n^2 + 6) - x_n = \frac{1}{5}(x_n^2 - 5x_n + 6) = \frac{1}{5}(x_n - 2)(x_n - 3)$$

Since $2 \leq x_n \leq 3$, the first factor is positive and the second negative, so that $x_{n+1} \leq x_n$ for all n. But (x_n) is also bounded below by 2, hence it must converge to some $x \in [2, 3]$. The limit satisfies $x = \lim_{n \to \infty} x_{n+1} = \frac{1}{5}(x^2 + 6)$, hence $(x - 2)(x - 3) = 0$. But $x \leq x_1 = 2.5$, since (x_n) decreases, so that $x = 2$.

2. (i) oscillates unboundedly: $x_{2n} = (-2n)^{2n} = (2n)^{2n} \to \infty$, and $x_{2n+1} = -(2n + 1)^{2n+1} \to -\infty$. So (x_n) oscillates unboundedly.
   (ii) $x_n = 2^n \to \infty$ as $n \to \infty$.
   (iii) $x_n = -2^n \to -\infty$ as $n \to \infty$.
   (iv) Though bounded below by 0, (x_n) oscillates unboundedly.

3. (i) To see that $\alpha = \underline{\lim}_n x_n$ is the *smallest* accumulation point of the sequence (x_n), note that for any $\varepsilon > 0$ we cannot construct a subsequence converging to $\gamma = \alpha - \varepsilon$, since for all except finitely many n, $x_n \geq \alpha = \gamma + \varepsilon$. Hence it suffices to show that α is an accumulation point. But by definition $x_n < \alpha + \varepsilon$ must hold for infinitely many n, while nearly all $x_n > \alpha - \varepsilon$, hence infinitely many $x_n \in N(\alpha, \varepsilon)$, which completes the proof.
   (ii) From the definitions, $\alpha - \varepsilon < x_n < \beta + \varepsilon$ except possibly for finitely many n.
   (iii) If (x_n) converges, its limit is the unique accumulation point, by Proposition 1, hence $\alpha = \beta$. Conversely, if $\alpha = \beta$ are finite, then (ii) above shows that $|x_n - \alpha| < \varepsilon$ for nearly all n, hence $x_n \to \alpha$. The cases when $\alpha = \pm\infty$ are similar.

4. (i) $\overline{\lim}_n x_n = 1$, since if $a > 1$ then $a > 1 + \frac{1}{n}$ for all $n > N$ for some $N \in \mathbb{N}$. Similarly $\underline{\lim}_n x_n = -1$. On the other hand $\inf_n x_n = x_1 = -2$, and $\sup_n x_n = x_2 = \frac{3}{2}$.
   (ii) We saw that for each $x \in [0, 1]$ some subsequence of (x_n) converges to x. On the other hand, $(x_n) \subset [0, 1]$. Hence $1 = \sup_n x_n = \overline{\lim}_n x_n$ and $0 = \inf_n x_n = \underline{\lim}_n x_n$, since the upper and lower limits are (resp.) the largest and smallest accumulation points of the sequence.

5. $a_{n+1} = \frac{1}{2+a_n}$ shows that $a_n \in [0, 1]$ implies $a_{n+1} \in [0, 1]$. Since $a_1 = 1$, it follows that $a_n \in [0, 1]$ for all n. Also:

$$a_{n+2} - a_n = \frac{1}{2 + a_{n+1}} - \frac{1}{2 + a_{n-1}} = \frac{-(a_{n+1} - a_{n-1})}{(2 + a_{n+1})(2 + a_{n-1})}$$

shows that $(a_{n+2} - a_n)$ and $(a_{n+1} - a_{n-1})$ have opposite signs. Setting $b_n = a_{2n-1}$ and $c_n = a_{2n}$, this means that (c_n) is increasing if and only if (b_n) is decreasing. Now note that $b_2 = a_3 = \frac{3}{7} < 1 = a_1 = b_1$, and that by the above, $(b_n - b_{n-1})$ and $(b_{n+1} - b_n)$ have the same sign. So (b_n) decreases and (c_n) increases. Both are bounded, hence $b = \lim_n b_n$ and $c = \lim_n c_n$ exist in $[0, 1]$. But for all n, $c_n = a_{2n} = \frac{1}{2+a_{2n-1}} = \frac{1}{2+b_n}$ and $b_{n+1} = a_{2n+1} = \frac{1}{2+a_{2n}} = \frac{1}{2+c_n}$. Hence $c = \frac{1}{2+b}$ and $b = \frac{1}{2+c}$. This yields $c^2 + 2c - 1 = 0$, and $c = \sqrt{2} - 1 = b$ is the only solution in $[0, 1]$, and is the limit of the sequence (a_n).

EXERCISES ON 3.2

1. In each case we use the Ratio Test:

$$\frac{(n+1)^{n+1}}{(n+1)!} \times \frac{n!}{n^n} = (1 + \frac{1}{n})^n \to e > 1$$

as $n \to \infty$, hence the series in (ii) diverges, and for the same reason that in (i) converges. For (iii) we calculate:

$$\frac{(n+1)!(n+5)!}{(2n+2)!} \times \frac{(2n)!}{n!(n+4)!} = \frac{(n+1)(n+5)}{(2n+1)(2n+2)} \to \frac{1}{4}$$

hence the series converges.

2. (i) This has the form $\sum_{k\geq1} \frac{1}{k^p}$ with $k = n - 5$ and $p = \frac{6}{5} > 1$. Hence the series converges.
    (ii) $\cos n$ lies (strictly) between -1 and 1, so the numerator is always positive and less than 2. The series converges by comparison with $2\sum_{n\geq1} \frac{1}{2^n} = \sum_{n\geq0} \frac{1}{2^n}$.
    (iii) $(2^n + 3^n)^{\frac{1}{n}} = 3\{(\frac{2}{3})^n + 1\}^{\frac{1}{n}}$ converges to 3, so the series is eventually term for term less than (e.g.) $4\sum_n \frac{1}{2^n}$, and thus converges.

EXERCISES ON 3.3

1. The case $r = 1$ is done in the text. For $r \neq 1$ we obtain

$$\int_1^n x^{-r}dx = \frac{1}{1-r}[n^{1-r} - 1]$$

which converges if and only if $r > 1$. Thus the same applies to the series $\sum_n n^{-r}$, by the Integral Test.

2. We need the first n such that $\gamma + \log_e n \geq 10$ (resp. 100, 1000). Taking $\gamma = 0.577$, we obtain $n \geq e^{9.423}$ (etc.) and our estimates become (i) 12 370, (ii) 1.5096×10^{43} (the estimate for (iii) is well beyond the capacity of many calculators).

EXERCISES ON 3.4

1. (i) $\log_n \leq n$, so the series converges absolutely by comparison with $\sum_n \frac{1}{n^2}$.
    (ii) Similar to (i) since $n + \sqrt{n} \leq 2n$.
    (iii) $|\sin x| \leq 1$ for all x, hence the series converges absolutely by comparison with $\sum_n \frac{1}{n^p}$ for $p = \frac{3}{2}$.
    (iv) The series does not converge absolutely: $\frac{n}{n^2+1} \geq \frac{1}{2n}$ for all n, so the series $\sum_n \frac{n}{n^2+1}$ diverges by comparison with the harmonic series. However, $\frac{n}{n^2+1} \leq \frac{1}{n}$ and so clearly decreases to 0, so that the alternating series test shows that $\sum_n (-1)^{n-1} \frac{n}{n^2+1}$ converges. So the series converges conditionally.
    (v) $a_n = \frac{2^n}{n^2}$ does not converge to 0, so the series diverges.
    (vi) Converges by the ratio test, hence absolutely.

2. In fact, $\sum_n a_n$ converges *absolutely* if and only if both $\sum_n a_n^+$ and $\sum_n a_n^-$ converge: note that the term-by-term sum of two convergent series is con-

vergent, and so is their term-by-term difference. But $|a_n| = a_n^+ + a_n^-$ and $a_n = a_n^+ - a_n^-$.

3. If $c_n = a_n + b_n$ and $\sum_n c_n$ and $\sum_n a_n$ are both convergent, then so is $\sum_n b_n$, since $b_n = c_n - a_n$. Hence if $\sum_n a_n$ converges and $\sum_n b_n$ diverges, then $\sum_n (a_n + b_n)$ diverges.

Chapter 4

EXERCISE ON 4.2

1. We have $c_n = \sum_{k=0}^{n} a_n b_{n-k}$, so that $c_0 = 1$, $c_1 = -(\frac{1}{\sqrt{2}} + \frac{1}{\sqrt{2}})$, $c_2 = \frac{1}{\sqrt{3}} + \frac{1}{\sqrt{2}}\frac{1}{\sqrt{2}} + \frac{1}{\sqrt{3}}$ etc. which gives the general term $c_n = (-1)^n \sum_{k=0}^{n} \frac{1}{\sqrt{(k+1)(n-k+1)}}$ as required. Now

$$(n-k+1)(k+1) = nk + n - k^2 + 1 = \left(\frac{n}{2}+1\right)^2 - \left(\frac{n}{2}-k\right)^2 \le \left(\frac{n}{2}+1\right)^2$$

so that $\sqrt{(n-k+1)(k+1)} \le 1 + \frac{n}{2} = \frac{n+2}{2}$. Hence $|c_n| \ge \sum_{k=0}^{n} \frac{2}{n+2} = 2(\frac{n+1}{n+2}) > 1$ for all $n > 1$. So (c_n) cannot converge to 0, hence the series $\sum_n c_n$ diverges, even though it is the Cauchy product of the convergent series $\sum_n (-1)^n \frac{1}{\sqrt{n+1}}$ with itself – as claimed in Tutorial Problem 4.2.

EXERCISES ON 4.3

1. (i) $R = 3$, (ii) $R = \frac{1}{2}$, (iii) $R = 1$, (iv) since $(\frac{1}{2} + \frac{1}{n})^n = \frac{1}{2^n}(1 + \frac{2}{n})^n$, we see that $R = \sqrt{2}$, (v) $R = 1$, (vi) $R = \frac{27}{4}$.
2. It is clear that $R = 1$, since r is fixed. This means that the series converges whenever $|x| < 1$, so that the sequence of its terms $(n^r x^n)_{n \ge 1}$ is null.

EXERCISE ON 4.4

1. (i) For $x > 0$, $\exp x = \sum_{n=0}^{\infty} \frac{x^n}{n!} > 1 + x$, hence $\exp x \to \infty$ as $x \to \infty$.
   (ii) Since $\exp x = \frac{1}{\exp(-x)}$ it follows from (i) that $\exp x \to 0$ as $x \to -\infty$.
   (iii) For $x > 0$, $\exp x > \frac{x^{n+1}}{(n+1)!}$, hence $\frac{\exp x}{x^n} > \frac{x}{(n+1)!}$, so that $x^n \exp(-x) < \frac{(n+1)!}{x}$ for each fixed n. Hence $\lim_{x \to \infty} x^n \exp(-x) = 0$.

Chapter 5

EXERCISE ON 5.2

1. The graphs are straightforward. Only (iv) is a one-one function.

EXERCISES ON 5.3

1. $f(x) = \frac{(1+x)^2 - 1}{x} = \frac{2x + x^2}{x} = 2 + x$ if $x \ne 0$. So $\lim_{x \to 0} f(x) = 2$ is the correct value needed at $x = 0$.
2. The 'gaps' occur at $x = \pm 2$. For $x \ne \pm 2$ we have

$$f(x) = \frac{(x-2)(x^2 + 2x + 4)}{(x-2)(x+2)} = x + \frac{4}{x+2}$$

Thus at $x = 2$ the value $f(2) = \lim_{x \to 2} f(x) = 3$ will remove the discontinuity, but at $x = -2$ the right- and left-limits do not coincide, so the 'gap' remains.

EXERCISES ON 5.4

1. By the sum, product and quotient rules (since $x^3 + 4x + 2 \neq 0$ in a neighbourhood of 2) we have $\lim_{x \to 2} f(x) = \frac{7}{18}$.
2. Clearly $\lim_{y \to 1} f(y) = 0$, since $f(y) = 0$ for all $y \neq 1$. But $f(g(x)) = f(1) = 2$ for all x, so $f \circ g$ cannot have limit 0 as $x \to a$ for any a. This example shows that we cannot interchange composition and limit operations without further restrictions (such as continuity).

Chapter 6

EXERCISES ON 6.1

1. The graph consists of two 'lines' intersecting at $x = \frac{1}{2}$. Suppose $a \neq \frac{1}{2}$. There exist sequences (x_n), (y_n) converging to a, such that each x_n is rational and each y_n is irrational. Thus $f(x_n) = x_n$ and $f(y_n) = 1 - y_n$ for each $n \in \mathbb{N}$. Now if $a \in \mathbb{Q}$ we have $f(a) = a$, so that $f(x_n) = x_n \to a = f(a)$, but $f(y_n) = 1 - y_n \to 1 - a \neq a = f(a)$, since $a \neq \frac{1}{2}$. On the other hand, if $a \notin \mathbb{Q}$ then $f(a) = 1 - a$, and $f(x_n) = x_n \to a \neq 1 - a$. Hence in both cases f is not continuous at $a \neq \frac{1}{2}$. On the other hand, if $a = \frac{1}{2}$ then $x_n \to \frac{1}{2}$ implies that $1 - x_n \to 1 - \frac{1}{2} = \frac{1}{2}$ also. Hence $f(x_n) \to \frac{1}{2}$ whenever $x_n \to \frac{1}{2}$. So f is continuous at $\frac{1}{2}$.
2. Fix $a \in \mathbb{R}$. Since each interval $(a - \frac{1}{n}, a + \frac{1}{n})$ contains at least one rational x_n it follows that there is a sequence (x_n) of rationals such that $x_n \to a$. Since f is continuous at a we must have $f(x_n) \to f(a)$ also. But $f(x_n) = 0$ for all n, hence $f(a) = 0$ also. This holds for each real a, so f is identically 0, as required.
3. $|\cos(\frac{1}{x})| \leq 1$ for all x, so $\lim_{x \to 0} f(x) = 0$, hence we need to set $f(0) = 0$ to obtain a continuous extension of f.

EXERCISES ON 6.2

1. (i) $f(0) = 0$ makes this function continuous at 0.
    (ii) There is a removable discontinuity at $x = 2$ (extend f by setting $f(2) = 3$ as in Exercise 2 in Section 5.3) and an unbounded jump at $x = -2$, since the left- and right-limits are $-\infty$ and $+\infty$ respectively.
    (iii) Jump discontinuities (in fact, at each $a = \sqrt{n}$ for $n \in \mathbb{N}$) with left-limits 1 and right-limits 0. The graph consists of ever 'straighter' pieces of the parabola.
2. f is *discontinuous at* $a \in \mathcal{D}_f$ if there exists a sequence $(x_n) \subset \mathcal{D}_f$ such that $x_n \to a$ as $n \to \infty$, but $(f(x_n))_n$ does not converge to $f(a)$.

EXERCISES ON 6.3

1. $f(2) = f(1).f(1) = a^2$, and by induction, $f(n + 1) = f(n).f(1) = a^n.a = a^{n+1}$ for all $n \in \mathbb{N}$. Clearly $0 < a = f(1) = f(1).f(0)$ so $f(0) = 1$. Hence $f(-n) = a^{-n}$ also. Now let $r = \frac{m}{n} \in \mathbb{Q}^+$ be given. Then $a^m = f(m) = f(nr) = \{f(r)\}^n$, so that $f(r) = a^r$ as claimed. The extension to all $r \in \mathbb{Q}$ is clear. If f is continuous on

$\mathbb{R}$, the real function g with $g(x) = f(x) - a^x$ is continuous on $\mathbb{R}$ and $g(r) = 0$ for all $r \in \mathbb{Q}$. Hence $g \equiv 0$, so $f(x) = a^x$ for all real x.

2. $x^a.x^b = e^{a \log x}.e^{b \log x} = e^{(a+b) \log x} = x^{a+b}$. Similarly, $(x^a)^b = y^b = e^{b \log y}$, where $y = x^a = e^{a \log x}$. But then $\log y = \log(\exp(a \log x)) = a \log x$, hence $y^b = e^{ab \log x} = x^{ab}$.

EXERCISE ON 6.4

1. Since $\exp x = 1 + x + \frac{x^2}{2!} + \ldots + \frac{x^n}{n!} + \ldots$, we obtain, for $x \neq 0$,

$$\frac{\exp(x) - 1}{x} = 1 + \frac{x}{2!} + \ldots + \frac{x^{n-1}}{n!} + \ldots,$$

and since this power series is continuous at $x = 0$ it follows that $\lim_{x \to 0} \frac{\exp x - 1}{x} = 1$. The argument for $\frac{\sin x}{x}$ is similar, since for $x \neq 0$, $\frac{\sin x}{x} = 1 - \frac{x^2}{3!} + \frac{x^4}{5!} - \ldots$

Chapter 7

EXERCISES ON 7.1

1. (i) For $x < 0$ this is just $|x|$, and for $x \geq 0$ it becomes $3x$. The function is clearly continuous on I. Hence $f(I) = [0, 6]$.
   (ii) $f(x) = \sin x - \cos x$ is continuous on I and $f(0) = -1$, $f(\frac{\pi}{2}) = 1$ are the extreme points. Hence $f(I) = [-1, 1]$.
2. The function f with $y = f(x) = x$ on $(0, \frac{1}{2}]$ and $y = f(x) = x + 1$ on $(\frac{1}{2}, 1)$ is clearly strictly monotone on $(0, 1)$ and has a jump discontinuity at $\frac{1}{2}$. Its inverse takes the values $x = f^{-1}(y) = y$ on $(0, \frac{1}{2}]$ and $x = f^{-1}(y) = y - 1$ on $(\frac{3}{2}, 2)$. The inverse is continuous at each point in its domain $\mathcal{D}_{f^{-1}} = \mathcal{R}_f = (0, \frac{1}{2}] \cup (\frac{3}{2}, 2)$, which is *not* an interval. This suggests that if f is strictly monotone on an open interval, then its inverse will always be continuous.

EXERCISES ON 7.2

1. Let $q(x) = \frac{P_n(x)}{x^n} - 1$, so that $P_n(x) = x^n(1 + q(x))$. As in Proposition 3, it is easy to see that $P_n(x)$ has the same sign as x^n for large enough x, (e.g. for $|x| > 1 + M$, where M is the largest of the coefficients $|a_k|$, $k = 0, 1, \ldots n - 1$). Since n is even, this means that $P_n(x) > 0$ whenever $|x|$ is large, i.e. we can find $x_1 < 0$ and $x_2 > 0$ such that $P_n(x_1) > 0$ and $P_n(x_2) > 0$. But $P_n(0) = a_0 < 0$, so P_n has a root in $(x_1, 0)$ and another root in $(0, x_2)$, by the *IVT*.
2. Since for $x > 0$ the series defining $\sin x$ has alternating signs, it follows that $|\sin(\frac{x-y}{2})| \leq \frac{|x-y|}{2}$ (in Chapter 9 this also follows from the Mean Value Theorem). Moreover $\sin 1 = \alpha < 1$, hence $|\cos x - \cos y| \leq \alpha |x - y|$. Thus by Proposition 4 the iteration $x_0 = 1$, $x_{n+1} = \cos x_n$ will yield $x = \lim_{n \to \infty} x_n$ as the unique solution of $x = \cos x$ in $[0, 1]$.
3. Since the maps $\cos$ and $x \mapsto x^4$ map $[0, 1]$ into itself, their composition f is a continuous function from $[0, 1]$ into $[0, 1]$, and thus has a fixed point, a in $[0, 1]$. As $\cos$ is positive on $[0, 1]$, we have $\cos a = \sqrt[4]{a}$.

EXERCISES ON 7.3

1. The function $f(x) = x - [x]$ has range $[0, 1)$, hence it is bounded. Since $f([0, 1]) = [0, 1)$, f does not have a maximum value on $[0, 1]$. The Boundedness Theorem, however, applies only to continuous functions, and f is not continuous at $x = 1$, since $\lim_{x \uparrow 1} f(x) = 1$ and $f(1) = 0$.

2. $g = \frac{1}{f}$ is well-defined and continuous on $[0, 1]$. Since the range $\mathcal{R}_f$ of f is a closed bounded interval (by the Boundedness Theorem) and does not contain 0, we may assume that $f([0, 1]) = [\alpha, \beta]$ with $\alpha > 0$. Then $g([0, 1]) = [\frac{1}{\beta}, \frac{1}{\alpha}]$ is also bounded.

3. Note that for all n, $0 \le f(x_n) \le r^{n-1} f(x_1) \le r^{n-1}$, since f maps $[0, 1]$ into itself. By the Bolzano–Weierstrass Theorem, the bounded sequence (x_n) has a convergent subsequence $(x_{n_k})_k$ with limit $c \in [0, 1]$. Since f is continuous, $f(c) = \lim_k f(x_{n_k})$. Thus $0 \le f(c) \le r^{n_k - 1}$ for all $k \ge 1$, and since $r < 1$ this means that $f(c) = 0$.

EXERCISE ON 7.4

1. The bounded function $f(x) = \sin(\frac{1}{x})$ is continuous on $(0,1)$, but $x_n = \frac{1}{(2n-\frac{1}{2})\pi}$ and $y_n = \frac{1}{(2n+\frac{1}{2})\pi}$ are points for which $f(x_n) = -1$, $f(y_n) = 1$, so that $|f(x_n) - f(y_n)| = 2$. On the other hand, for large enough n, $|x_n - y_n| = \frac{2}{\pi}(\frac{1}{4n-1} - \frac{1}{4n+1})$ is less than any preassigned $\delta > 0$. Thus for $\varepsilon < 2$ no $\delta > 0$ can be found to satisfy the definition of uniform continuity. This shows why we restrict to functions on *closed* bounded intervals in Theorem 4.

Chapter 8

EXERCISES ON 8.2

1. Let $x \in N(a, \delta) \setminus \{a\}$, where $\delta > 0$ is chosen so that $0 \notin N(a, \delta)$. Then

$$\frac{f(x) - f(a)}{x - a} = \frac{\frac{1}{x} - \frac{1}{a}}{x - a} = \frac{-(x - a)}{ax(x - a)} = -\frac{1}{ax} \to -\frac{1}{a^2} \text{ as } x \to a$$

Thus $f'(a) = -\frac{1}{a^2}$ for all $a \ne 0$. Defining $f(0) = 0$ will not make f differentiable at 0, since the extended function is not even continuous at 0.

2. The function $f : x \mapsto x - [x]$ is discontinuous at each $n \in \mathbb{Z}$, as we have seen before. When $x \in (n, n + 1)$ we have $f(x) = x - n$, which has derivative 1 at each such x. Hence $f'(x) = 1$ and f' has domain $\mathbb{R} \setminus \mathbb{Z}$.

EXERCISES ON 8.3

1. If $f : \mathbb{R} \mapsto \mathbb{R}$ and a is an interior point of $\mathcal{D}_f$, then the function f' is defined at a if $L = \lim_{x \to a} \frac{f(x) - f(a)}{x - a}$ exists, and the value of f' is then $f'(a) = L$. Now suppose that a is also an interior point of the domain of f', and that the limit $\lim_{x \to a} \frac{f'(x) - f'(a)}{x - a} = l$ exists. Then l is said to be the value of the second derivative f'' of f at a, and $f'' : \mathbb{R} \mapsto \mathbb{R}$ has domain $\mathcal{D}_{f''}$ equal to the set of points where this limit exists.

2. $P(1) = 0$, so $(x - 1)$ is a factor of $P(x)$. Hence $P(x) = (x - 1)Q(x)$ for some (6th degree) polynomial Q. Now $P'(x) = 7x^6 - 18x^5 + 15x^4 - 3x^2 - 6x + 5$, so

$P'(1) = 0$. On the other hand, $P'(x) = Q(x) + (x - 1)Q'(x)$ (borrowing the product rule for now!) so $P'(1) = Q(1)$. Hence $(x - 1)$ is also a factor of Q, i.e. $P(x) = (x - 1)^2 R(x)$ for some polynomial R. Again, since $Q(x) = (x - 1)R(x)$ we have $Q'(x) = R(x) + (x - 1)R'(x)$, so that $Q'(1) = R(1)$. On the other hand, $P''(x) = 2Q'(x) + (x - 1)Q''(x)$, hence $P''(1) = R(1)$. But $P''(x) = 42x^5 - 90x^4 + 60x^3 - 6x - 6$, so $P''(1) = 0$ also, and hence $(x - 1)$ is a factor of R. Hence we have shown that $P(x) = (x - 1)^3 S(x)$ for some polynomial S.

3. $\frac{f(a+h)-f(a-h)}{2h} = \frac{f(a+h)-f(a)+f(a)-f(a-h)}{2h} = \frac{1}{2}\{\frac{f(a+h)-f(a)}{h} + \frac{f(a)-f(a-h)}{h}\}$ and when we let $h \downarrow 0$, the first term on the right is the right-limit and the second the left-limit of f at a. When f is differentiable at a, these limits exist and both equal $f'(a)$. Hence $\lim_{h\downarrow 0}\frac{f(a+h)-f(a-h)}{h} = \frac{1}{2}\{f'(a) + f'(a)\} = f'(a)$ whenever f is differentiable at a. Considering $f(x) = |x|$ at $a = 0$ we have $\frac{|h|-|-h|}{2h} = 0$ for all $h > 0$, so the limit exists, even though f is not differentiable at 0. (This limit picks out the 'average' of the slopes of would-be 'tangents' to f at 0.)

EXERCISES ON 8.4

1. $f(x) = x^x = \exp(x \log x)$ for $x > 0$, so that by the chain and product rules, and properties of exp and log we have:

$$f'(x) = \exp(x \log x)\left(x.\frac{1}{x} + \log x\right) = x^x(1 + \log x)$$

For this to be 0, we need $1 + \log x = 0$, since exp is always strictly positive. Hence $x = \frac{1}{e}$ is the only solution. For $g(x) = x^{(x^x)}$, which makes sense for $x > 0$, and also at $x = 0$ if we define $0^0 = 1$, we obtain $g(x) = \exp(x^x \log x)$, so that

$$g'(x) = \exp(x^x \log x)\left[x^x\{1 + \log x\} \log x + x^x \frac{1}{x}\right]$$

$$= x^{(x^x)}\left[x^x\left\{\frac{1}{x} + \log x + (\log x)^2\right\}\right]$$

2. Nowhere: the function is defined only at ± 1.
3. Let $x \in (a, b)$ be given. If $g = f^{-1}$ were differentiable at $y = f(x)$ then we would be able to use the chain rule on the function $h = g \circ f$, since both f and g would then be differentiable (at y and x respectively). So we would have $h'(x) = g'(y).f'(x) = 0$, since $f'(x) = 0$ by assumption. But this is nonsense, since $h = g \circ f = f^{-1} \circ f$ is the identity map $x \mapsto x$, which has constant derivative $x \mapsto 1$. So $h'(x) = 1$, i.e. $0 = 1$. This contradiction is caused by the assumption that f^{-1} is differentiable at y, and this assumption is therefore false, i.e. f^{-1} is *not* differentiable at y.
4. $\sinh x = \frac{e^x - e^{-x}}{2}$ so that its derivative is easily seen to be $\cosh x = \frac{e^x + e^{-x}}{2}$. See *Calculus and ODEs* Chapter 7, Exercise 11 for the calculation of $\frac{d}{dx}\sinh^{-1} x = \frac{1}{\sqrt{1+x^2}}$ and $\frac{d}{dx}\cosh^{-1} x = \frac{1}{\sqrt{x^2-1}}$.

EXERCISES ON 8.5

1. By Example 4 in Chapter 9, $f'(x) = |x - 1|\{2x + \frac{x^2}{x-1}\}$ is well-defined at all points except $x = 1$. (Note in particular that f is differentiable at 0.)

2. Since $f(2) = -1$, we obtain $\frac{2a+b}{-2} = -1$, or $2a + b = 2$. Also $f'(2) = 0$, so by the quotient rule

$$f'(x) = \frac{a(x^2 - 5x + 4) - (ax + b)(2x - 5)}{(x^2 - 5x + 4)^2}$$

yields $-2a - (2a + b)(-1) = 0$, i.e. $4a + b = 0$. So $a = -1, b = 4$.

3. For $x + b \neq n\pi$, $f(x) = \frac{\sin(x+a)}{\sin(x+b)}$ is well-defined and has derivative $f'(x) = \frac{\cos(x+a)\sin(x+b) - \sin(x+a)\cos(x+b)}{\sin^2(x+b)} = \frac{\sin(a-b)}{\sin^2(x+b)}$, and this is never 0 unless $a = b + n\pi$ for some $n \in \mathbb{Z}$. But in the latter case $f(x) = 1$ for all $x \in \mathbb{R}$. So f is either constant or has no extrema. (The sketch is omitted.)

Chapter 9

EXERCISES ON 9.1

1. Part (iv) will suffice here: fix $x < y$ in $[a, b]$ and apply the *MVT* to find $c \in (x, y)$ with $\frac{f(y)-f(x)}{y-x} = f'(c)$. Since $f'(c) \leq 0$, and $y - x > 0$, we have $f(y) - f(x) \leq 0$, so f is decreasing on $[a, b]$.

2. Let $f(x) = 1 - x^p - p(1 - x)$ for $x > 0$. Then $f'(x) = -px^{p-1} + p = p(1 - x^{p-1})$, so $f'(1) = 0$, and since $p > 1, f' > 0$ on $(0, 1)$. So f is strictly increasing on $[0, 1]$, i.e. if $x \in (0, 1)$ then $f(x) < f(1) = 1 - 1^p - p(1 - 1) = 0$. Hence $1 - x^p < p(1 - x)$ for $x \in (0, 1)$.

3. $g(x) = f(x) - \gamma x$ is differentiable, hence continuous on $[a, b]$. Moreover, $g'(b) = f'(b) - \gamma > 0$ and $g'(a) = f'(a) - \gamma < 0$, hence a, b are not extreme points of g. On the other hand, g is continuous on $[a, b]$, so it must attain its minimum at some point c of $[a, b]$. As c cannot equal a or b, we have $a < c < b$ and $g'(c) = 0$. Hence $f'(c) = \gamma$.

EXERCISES ON 9.2

1. (i) $f(x) = \cos x + |\sin x|$ has $f'(x) = -\sin x + \frac{\sin x}{|\sin x|}\cos x$ except when $x = n\pi$ ($n \in \mathbb{Z}$). Maxima occur at $2n\pi \pm \frac{\pi}{4}$ for $n \in \mathbb{Z}$ (and are global, with value $\sqrt{2}$) and local minima at $2n\pi$ (value 1) with global minima (value -1) at $(2n + 1)\pi$. The function increases on $(0, \frac{\pi}{4}) + 2n\pi$ and $(-\pi, -\frac{\pi}{4}) + 2n\pi$ and decreases on the complementary open intervals.

(ii) $f(x) = |\sin x + \cos x|$ is defined on $\mathbb{R}$, and differentiable except when $\sin x + \cos x = 0$. So first consider the differentiable function $g(x) = \sin x + \cos x$, which has derivative $g' : x \mapsto \cos x - \sin x$, which is 0 iff $x = k\pi + \frac{\pi}{4}$ for some integer k. Also, $g' > 0$ on $(-\frac{3\pi}{4}, \frac{\pi}{4}) + 2k\pi$ and $g' < 0$ on $(\frac{\pi}{4}, \frac{5\pi}{4}) + k\pi$ for $k \in \mathbb{Z}$.

Thus the strict local minimum points are at $-\frac{3\pi}{4} + 2k\pi$, and the strict local maxima are at $\frac{\pi}{4} + 2k\pi$. However, $f = |g|$ will have strict local maxima at all these extreme points of g, and its minimum value 0 will be attained when $g(x) = 0$, i.e. at $x = \frac{3\pi}{4} + k\pi$. This yields the graph of f. Note that f is differentiable at each maximum point, but at none of the minimum points!

(iii) is similar to the above: f increases on intervals of the form $[0, \frac{\pi}{4}] + \frac{k\pi}{2}$, is not differentiable at points $\frac{k\pi}{2}$, and has range $[1, \sqrt{2}]$.

2. (i) $f(x) = |x^2 - x^4| = x^2|1 - x^2|$ is differentiable unless $x = \pm 1$. Here $\mathcal{D}_f = \mathbb{R}$ and $\mathcal{R}_f = [0, \infty)$, while $f'(x) = \frac{x^2 - x^4}{|x^2 - x^4|}(2x - 4x^3)$ so that $0, \pm 1, \pm\frac{1}{\sqrt{2}}$ are the only possible extreme points. The first three are (global and) strict local minima, and $\pm\frac{1}{\sqrt{2}}$ are strict local maxima, but not global.

(ii) $f(x) = |x|e^{-|x|}$ is positive, even, and has $\mathcal{D}_f = \mathbb{R}$, $\mathcal{R}_f = (0, \frac{1}{e}]$, since the maximum occurs if $-xe^{-x} + e^{-x} = 0$, i.e. at ± 1. f is not differentiable at 0, since $f'_+(0) = 1 \neq -1 = f'_-(0)$. The strict minimum at 0 is global, and the maxima at $\pm\frac{1}{\sqrt{2}}$ are global (and strict local).

(iii) $f(x) = x^x$ has $\mathcal{D}_f = [0, \infty)$ if we set $0^0 = 1$. It is differentiable for $x > 0$ and $f'(x) = x^x(1 + \log x)$, so the only (strict) minimum occurs at $\frac{1}{e}$. Thus $\mathcal{R}_f = [e^{-\frac{1}{e}}, \infty)$, and there is a strict local maximum at 0.

3. For $c \neq 0, f(x) = \frac{ax+b}{cx+d}$ is defined and differentiable unless $x = -\frac{d}{c}$ and then

$$f(x) = \frac{a(cx + d) - c(ax + b)}{(cx + d)^2}$$

which is 0 iff $ad = bc$, i.e. either $b = 0$ or $\frac{a}{c} = \frac{b}{d} = r$, say. If $b = 0$ and $a = 0$, then $f \equiv 0$, If $b = 0$ and $a \neq 0$, then $d = 0$, so that $f(x) = \frac{ax}{bx} = \frac{b}{d}$, and f is again constant. If $b \neq 0$, $ax + b = r(cx + d)$ and again $f(x) = r$ for all x. Hence in all cases f is constant and therefore has no (non-trivial) extreme points. (The graph is obvious.)

EXERCISES ON 9.3

1. (i) By l'Hôpital's rules: $\lim_{x\to 0}(\frac{1}{\sin x} - \frac{1}{x}) = \lim_{x\to 0}(\frac{x - \sin x}{x \sin x}) = \lim_{x\to 0}(\frac{1 - \cos x}{\sin x + x \cos x}) = \lim_{x\to 0}(\frac{\sin x}{2 \cos x - x \sin x}) = 0$.

(ii) $|\cos(\frac{1}{x})| \leq 1$ for all x, hence $\lim_{x\to 0} x \cos(\frac{1}{x}) = 0$.

(iii) $\lim_{x\to 0} \frac{1 - \cos x}{x^2} = \lim_{x\to 0} \frac{\sin x}{2x} = \lim_{x\to 0} \frac{\cos x}{2} = 2$ by l'Hôpital's rules.

(iv) Similarly, $\lim_{x\to 0} \frac{\log(1 + tx)}{x} = \lim_{x\to 0} \frac{\frac{t}{1+tx}}{1} = t$. Hence with $t = 1$ and using the null sequence $x_n = \frac{1}{n}$ we have $1 = \lim_{n\to\infty} \frac{\log(1 + \frac{1}{n})}{\frac{1}{n}} = \lim_{n\to\infty} n \log(1 + \frac{1}{n}) = \lim_{n\to\infty} \log(1 + \frac{1}{n})^n$. Since exp is continuous, we obtain $e = \lim_{n\to\infty}(1 + \frac{1}{n})^n$.

2. Using (iv) with $\frac{1}{t}$ replacing x and x replacing t, and the continuity of exp provides $\lim_{t\to\infty}(1 + \frac{x}{t})^t = \exp(x)$.

3. We know that $f(x) = x^2 \cos(\frac{1}{x})$ is differentiable at 0 if we set $f(0) = 0$. Then also $f'(0) = 0$. But although $\frac{f(x)}{g(x)} = x \cos(\frac{1}{x})$ has limit 0 at 0, the right-hand side does not make sense: we have $\frac{f'(x)}{g'(x)} = \frac{2x \cos(\frac{1}{x}) - \sin(\frac{1}{x})}{1}$ for $x \neq 0$, but this does not have a limit at $x = 0$.

EXERCISES ON 9.4

1. The case $n = 1$ is the product rule. Assuming the result for $n = k - 1$, we need to verify it for $n = k$. This is notationally tedious, but relies simply on the identity

$\binom{k-1}{r-1} + \binom{k-1}{r} = \binom{k}{r}$. The details are omitted. Applying the formula to $f(x) = x^4$, $g(x) = \cos x$ we obtain

$$h^{(4)}(x) = \sum_{r=0}^{4} \binom{4}{r} f^{(r)}(x) g^{(n-r)}(x)$$

$$= 24 \cos x - 96x \sin x - 72x^2 \cos x + 12x^3 \sin x + x^4 \cos x$$

2. $f(x) = \cosh^2 x - \sinh^2 x$ is differentiable and $f'(x) = 2 \cosh x \sinh x - 2 \sinh x \cosh x = 0$ for all x. Hence f is constant. But $f(0) = 1$, so $\cosh^2 x - \sinh^2 x = 1$.

EXERCISES ON 9.5

1. Taylor's expansion of order $(m-1)$ for f about a becomes $f(x) - f(a) = \frac{f^{(m)}(c)}{m!}(x-a)^m$, since $f^{(k)}(a) = 0$ for $k < m$. Now $f^{(m)}$ remains non-zero throughout some neighbourhood N of a (since it is a continuous function) hence its sign does not change there. If m is even $(x-a)^m$ remains positive, so the sign of $f^{(m)}(c)$ determines everything: if it is negative, $f(x) < f(a)$ throughout N, and f has a maximum at a; if it is positive, f has a minimum. If m is odd, the sign of $(x-a)^m$ changes as we move through N, hence so does that of $(f(x) - f(a))$ and there is no extremum at a.

$f(x) = x^6 - x^4$ has $f^{(4)}(0) = -24$ as the first non-zero derivative at 0, hence has a local maximum there. Since $f'(x) = 6x^5 - 4x^3$ reduces to $6x^2 - 4$ for $x \neq 0$, the other critical points are $\pm\sqrt{\frac{2}{3}}$. It should be clear that these are strict local minimum points: on $(0, \sqrt{\frac{2}{3}})$ $6x^5 < 4x^3$, and the inequality reverses on $(-\sqrt{\frac{2}{3}}, 0)$.

2. Compute the polynomials for $\log(1 + x)$ and $\log(1 - x)$ separately:

$$\log(1 + x) = x - \frac{x^2}{2} + \frac{x^3}{3} - \ldots + \frac{x^9}{9} + \frac{x^{10}}{10!} f_1^{(10)}(c_1)$$

$$\log(1 - x) = -x - \frac{x^2}{2} - \frac{x^3}{3} - \ldots - \frac{x^9}{9} + \frac{x^{10}}{10!} f_2^{(10)}(c_2)$$

where $f_1(x) = \log(1 + x)$, $f_2(x) = \log(1 - x)$ and c_i ($i = 1, 2$) are between 0 and x. Thus we obtain $\log(\frac{1+x}{1-x}) = 2(x + \frac{x^3}{3} + \frac{x^5}{5} + \frac{x^7}{7} + \frac{x^9}{9}) + E(x)$ where $E(x) = \frac{x^{10}}{10!} 9! \{\frac{1}{(1-c_2)^{10}} - \frac{1}{(1-c_1)^{10}}\}$ is the error term. An estimate for $\log 3$ is obtained by setting $x = \frac{1}{2}$, so that $\log 3 = \log(\frac{1+x}{1-x}) = 2(2^{-1} + \frac{1}{3}2^{-3} + \frac{1}{5}2^{-5} + \frac{1}{7}2^{-7} + \frac{1}{9}2^{-9})$, and this is accurate to within $|E(\frac{1}{2})|$.

3. Let $f(x) = \log(\sec x)$, and expand at $a = 0$: then $f(0) = 0$, and

$$f'(x) = \tan x \qquad\qquad f'(0) = 0$$
$$f''(x) = \sec^2 x \qquad\qquad f''(0) = 1$$
$$f'''(x) = 2 \sec^2 x \tan x \qquad\qquad f'''(0) = 0$$
$$f''''(x) = 6 \sec^4 x - 4 \sec^2 x$$

Hence the third degree Taylor polynomial of f is simply $T_3(x) = \frac{x^2}{2}$, while the remainder is $R_4(x) = \frac{x^4}{4!} f''''(c)$ for some c between 0 and x. Consider this with

$x \in (-\frac{\pi}{4}, \frac{\pi}{4})$, so that $\sec^2 x$ varies between 1 and 2, so that $f''''(c)$ lies between 2 and 16. This yields the inequality

$$x^4 \le \log(\sec x) - \frac{x^2}{2} \le 8x^4$$

which we have proved to be valid whenever $x \in (-\frac{\pi}{4}, \frac{\pi}{4})$.

Chapter 10

EXERCISES ON 10.2

1. Take $P_n = \{\frac{k}{n} : k = 0, 1, \ldots n\}$ as a partition of $[0, 1]$. Since f is increasing, $M_k = f(\frac{k}{n}) = \frac{k^3}{n^3}$ and $m_k = \frac{(k-1)^3}{n^3}$ for $k \le n$. Hence $\int_0^1 \psi_{P_n}^f = \frac{1}{n^4} \sum_{k=1}^n k^3 = \frac{n^2(n+1)^2}{4n^4}$ and $\int_0^1 \phi_{P_n}^f = \frac{1}{n^4} \sum_{k=1}^{n-1} k^3 = \frac{n^2(n-1)^2}{4n^4}$, both of which converge to $\frac{1}{4}$ when $n \to \infty$. Hence f is integrable and $\int_0^1 x^3 \, \mathrm{d}x = \frac{1}{4}$.

2. For $h = -f$, $\sup\{f(x) : x \in I]\} = -\inf\{h(x) : x \in I\}$ for every interval I. Apply this to each subinterval of the partition P to see that $U_P = \int_a^b \psi_P^f = -\int_a^b \phi_P^h$. The other case is proved similarly. Hence as $\overline{\int_a^b} f = \inf_P U_P = \sup_P L_P = \underline{\int_a^b} f$, it follows that h is integrable and $\int_a^b h = -\int_a^b f$, because $\underline{\int_a^b} h = \sup_P(-U_P) = -\inf_P U_P = -\sup_P L_P = \inf_P(-L_P) = \overline{\int_a^b} h$.

EXERCISES ON 10.3

1. First we show that the integral is linear: for any partition P we have

$$U(P, \alpha f + \beta g) \le \alpha U(P, f) + \beta U(P, f)$$

and

$$L(P, \alpha f + \beta g) \ge \alpha L(P, f) + \beta L(P, f)$$

since $\sup(\alpha f(x) + \beta g(x)) \le \alpha \sup f(x) + \beta \sup g(x)$, and the opposite inequality holds for inf. Since we can choose our partition to make $U(P, f) - L(P, f) < \varepsilon$ and similarly for g, it follows that Riemann's criterion is satisfied by $\alpha f + \beta g$ and that $\int_a^b (\alpha f + \beta g) = \alpha \int_a^b f + \beta \int_a^b g$.

   It is also clear that if $f \ge 0$ on $[a, b]$, then $L(P, f) \ge 0$ for any partition, and so $\int_a^b f \ge 0$. Thus, if $f \le g$, $g - f \ge 0$ implies $\int_a^b (g - f) \ge 0$, and by linearity this means that $\int_a^b f \le \int_a^b g$.

2. As in the Hint, $F(x) - F(y) = \int_a^x f - \int_a^y f = \int_y^x f = \int_y^{y+h} f$ and the 'area property' means that $|\int_y^{y+h} f| \le Mh$. The result follows.

EXERCISES ON 10.4

1. (i) f is continuous (and bounded) on $[-1, 1]$. $\int_{-1}^1 |x^3| \, \mathrm{d}x = 2 \int_0^1 x^3 \, \mathrm{d}x = \frac{1}{2}$.
   (ii) f is bounded and is constant on each interval $[-2, 0)$ and $(0, 2]$, so it is 'nearly' a step function. The only discontinuity is at $x = 0$. Since f is odd, $\int_{-2}^2 f(x) \, \mathrm{d}x = 0$.

(iii) This is a step function, hence integrable. It increases by 1 at each $\sqrt[3]{n}$ $n = 1, 2, \ldots, 8$ and the integral equals the sum of the areas of the rectangles involved.

2. $\int_0^3 f = \int_0^1 1 \, dx + \int_1^2 (x-1)^2 dx + \int_2^3 \sqrt{x-2} \, dx = 1 + \frac{1}{3} + \frac{2}{3} = 2.$

EXERCISES ON 10.5

1. By the linearity of the integral and earlier examples:

$$\int_0^1 (3x^3 - 2x^2 + 5\sqrt{x} - 1) \, dx$$

$$= 3 \int_0^1 x^3 \, dx - 2 \int_0^1 x^2 \, dx + 5 \int_0^1 \sqrt{x} \, dx - \int_0^1 1 \, dx$$

$$= \frac{3}{4} - \frac{2}{3} + \frac{10}{3} - 1 = \frac{29}{12}$$

2. $f_n(x) = \frac{\cos nx}{n^2 + x^2}$ is continuous on $[0, 1]$, hence integrable over $[0, 1]$. Moreover, $f_n \to 0$ uniformly on $[0, 1]$, since $|f_n(x)| \leq \frac{1}{n^2}$ for all $x \in [0, 1]$. Hence by Theorem 3, $\int_0^1 f_n \to 0$ as $n \to \infty$.

Chapter 11

EXERCISES ON 11.1

1. $\int_a^b f$ is the area bounded by the graph of f, the x-axis, and the vertical lines $x = a$, $x = b$. Similarly for g. Thus if $f \geq g$, the area between the graphs (and bounded by the same vertical lines) is $\int_a^b f - \int_a^b g = \int_a^b (f - g)$. Applying this with $f = \cos$, $g = \sin$ on the interval $[0, \frac{\pi}{4}]$ and $[\frac{5\pi}{4}, 2\pi]$, and with roles reversed on $[\frac{\pi}{4}, \frac{5\pi}{4}]$, we obtain the area between the graphs of sin and cos as:

$$\left(\int_0^{\frac{\pi}{4}} + \int_{\frac{5\pi}{4}}^{2\pi} \right)(\cos x - \sin x) \, dx + \int_{\frac{\pi}{4}}^{\frac{5\pi}{4}} (\sin x - \cos x) \, dx = 4\sqrt{2}$$

2. f is unbounded at 0, so that the Riemann integral of f is undefined over any interval which includes 0. The Fundamental Theorem cannot therefore be applied directly.

EXERCISES ON 11.2

1. Use $f(x) = x$, $g(x) = \log x$ in Proposition 1. Then

$$\int_a^b 1 . \log x \, dx = [x . \log x]_a^b - \int_a^b x . \frac{1}{x} dx$$

whenever f, g are integrable on $[a, b]$. This holds when $a = e^{-1}$, $b = e$, so $\int_{e^{-1}}^e \log x \, dx = [x . \log x]_{e^{-1}}^e - \int_{e^{-1}}^e dx = \frac{2}{e}$.

2. Use the change of variable $x = t^n$ and apply Proposition 2.

3. Since $x^2 + y^2 = 1$ is the equation of the unit circle, the integral $\int_0^1 \sqrt{1 - x^2} \, dx$ provides the area of the first quadrant (quarter-circle). The substitution

$x = \cos t$ yields

$$\int_0^1 \sqrt{1 - x^2}\, dx = \int_{\frac{\pi}{2}}^0 - \sin t \sqrt{1 - \cos^2 t}\, dt = \int_0^{\frac{\pi}{2}} \sin^2 t\, dt$$

using the fact that $\cos \frac{\pi}{2} = 0$ by our definition of π, and that $\cos 0 = 1$. The last integral is $\frac{\pi}{4}$, and this coincides with the area of the quarter-circle when we use the geometric definition of π. Thus the definitions are the same.

EXERCISES ON 11.3

1. (i) We find that $\int_0^1 \log x\, dx = \lim_{a \downarrow 0} \int_a^1 \log x\, dx = \lim_{a \downarrow 0} [x \log x - x]_a^1 = -1$, since $a \log a \to 0$ as $a \downarrow 0$.
   (ii) Here we split the integral, as the integrand is undefined at $x = \pm 1$. So $\int_{-1}^1 \frac{dx}{\sqrt{1-x^2}} = \lim_{a \downarrow -1} \int_a^0 \frac{dx}{\sqrt{1-x^2}} + \lim_{b \uparrow 1} \int_0^b \frac{dx}{\sqrt{1-x^2}}$. In each case $\sin^{-1}$ is a primitive of $\frac{1}{\sqrt{1-x^2}}$, the limits exist and their sum is π.
   (iii) Again we split at 0, and calculate $\int_{-b}^0 \frac{dx}{e^x + e^{-x}}$ and $\int_0^a \frac{dx}{e^x + e^{-x}}$ separately. Each integrand is $\frac{e^x}{e^{2x}+1}$, which becomes $\frac{1}{u^2+1}$ with the change of variable $u = e^x$, so the integrals are $[\tan^{-1} e^x]_{-b}^0$ and $[\tan^{-1} e^x]_0^a$, each of which has limit $\frac{\pi}{2}$, and so $\int_{-\infty}^\infty \frac{dx}{e^x+e^{-x}} = \pi$.

2. $f(x) = \frac{1}{x(x-1)} = \frac{1}{x-1} - \frac{1}{x}$ is continuous on each of the intervals $(-\infty, 0)$, $(0, 1)$ and $(1, \infty)$, and we need a, b to lie in the same interval in each case for f to be integrable over $[a, b]$. $F(x) = \log|x| - \log|x - 1| = \log|\frac{x}{x-1}|$ is a primitive in each case, hence $\int_a^b f = \log|\frac{b}{b-1}| - \log|\frac{a}{a-1}|$.

3. We have $\int_a^b (\alpha f + \beta g) = \alpha \int_a^b f + \beta \int_a^b g$ for all finite b, so by the algebra of limits the result follows.

EXERCISES ON 11.4

1. *Comparison Theorem:* Suppose f, g are defined on $(a, b]$ and Riemann-integrable on $[a + \varepsilon, b]$ whenever $0 < \varepsilon < b - a$. If $0 \le f \le g$ on $(a, b]$ and the improper integral $\int_a^b g$ converges, then so does $\int_a^b f$.
   Proof: For every $\varepsilon > 0$ $\int_{a+\varepsilon}^b f \le \int_{a+\varepsilon}^b g$, and since $\int_a^b g = \lim_{\varepsilon \downarrow 0} \int_{a+\varepsilon}^b g$ exists, the integrals $\int_{a+\varepsilon}^b g$ are bounded, i.e contained in some bounded interval. But $\int_{a+\varepsilon}^b f$ is increasing as $\varepsilon \downarrow 0$, hence the limit $\int_a^b f$ exists.

2. See Exercises on 3.3(1).

3. If $\int_a^\infty |f|$ converges, then $I = \int_a^\infty |f| = \lim_{n \to \infty} \int_a^n |f|$. Let $I_n = \int_a^n |f|$ and similarly define $J_n = \int_a^n f$. Then we can copy the proof of Proposition 3.5 to deduce that (J_n) converges (using $|\int_k^{k+1} f| \le \int_k^{k+1} |f|$ repeatedly), and its limit defines $J = \int_a^\infty f$. By the Comparison Test for integrals we then have $J_n \le I_n \le I$ for all $n \ge 1$, and similarly for $-J_n$. Hence $|J| \le I$, as required.

Index

Lightning Source UK Ltd.
Milton Keynes UK
03 October 2009

144411UK00001B/34/A